T0295168

Lean Six Sigma 4.0 for Operational Excellence Under the Industry 4.0 Transformation

This book presents innovative breakthroughs in operational excellence that can solve the operational issues of smart factories. It illustrates various tools and techniques of Lean Six Sigma 4.0 and details their suitability for manufacturing and service systems.

Lean Six Sigma 4.0 for Operational Excellence Under the Industry 4.0 Transformation provides technological advancement in operational excellence and offers a framework to integrate Lean Six Sigma and Industry 4.0. The book is a guide to dealing with new operational challenges and explains how to use Industrial IoT, Sensors, and AI to collect real-time data on the shop floor. While focusing on developing a toolset for Lean Six Sigma 4.0, this book also presents the enabling factors to adopt Lean Six Sigma 4.0 in the manufacturing and service sectors.

The book will help industrial managers, practitioners, and researchers on the path of process improvement in modern-day industries.

Sustainable Manufacturing Technologies: Additive, Subtractive, and Hybrid

Series Editors: Chander Prakash, Sunpreet Singh, Seeram Ramakrishna, and Linda Yongling Wu

This book series offers the reader comprehensive insights from recent research breakthroughs in additive, subtractive, and hybrid technologies while emphasizing their sustainability aspects. Sustainability has become an integral part of all manufacturing enterprises in order to provide various techno-social pathways toward developing environmentally friendly manufacturing practices. It has also been found that numerous manufacturing firms are still reluctant to upgrade their conventional practices to sophisticated sustainable approaches. Therefore, this new book series aims to provide a globalized platform to share innovative manufacturing mythologies and technologies. The books will explore the emerging issues of the conventional and non-conventional manufacturing technologies and cover recent innovations.

Advances in Manufacturing Technology: Computational Materials Processing and Characterization
Edited by Rupinder Singh, Sukhdeep Singh Dhami, and B. S. Pabla

Additive Manufacturing for Plastic Recycling: *Efforts in Boosting A Circular Economy*
Edited by Rupinder Singh and Ranvijay Kumar

Additive Manufacturing Processes in Biomedical Engineering: Advanced Fabrication Methods and Rapid Tooling Techniques
Edited by Atul Babbar, Ankit Sharma, Vivek Jain, Dheeraj Gupta

Additive Manufacturing of Polymers for Tissue Engineering: Fundamentals, Applications, and Future Advancements
Edited by Atul Babbar, Ranvijay Kumar, Vikas Dhawan, Nishant Ranjan, Ankit Sharma

Sustainable Advanced Manufacturing and Materials Processing: Methods and Technologies
Edited by Sarbjeet Kaushal, Ishbir Singh, Satnam Singh, and Ankit Gupta

3D Printing of Sensors, Actuators, and Antennas for Low-Cost Product Manufacturing
Edited by Rupinder Singh, Balwinder Singh Dhaliwal, and Shyam Sundar Pattnaik

Lean Six Sigma 4.0 for Operational Excellence Under the Industry 4.0 Transformation
Edited by Rajeev Rathi, Jose Arturo Garza-Reyes, Mahender Singh Kaswan, and Mahipal Singh

Handbook of Post-Processing in Additive Manufacturing: Requirements, Theories, and Methods
Edited by Gurminder Singh, Ranvijay Kumar, Kamalpreet Sandhu, Eujin Pei, and Sunpreet Singh

For more information on this series, please visit: www.routledge.com/Sustainable-Manufacturing-Technologies-Additive-Subtractive-and-Hybrid/book-series/CRCSMTASH

Lean Six Sigma 4.0 for Operational Excellence Under the Industry 4.0 Transformation

Edited by
Rajeev Rathi, Jose Arturo Garza-Reyes,
Mahender Singh Kaswan, and Mahipal Singh

CRC Press
Taylor & Francis Group
Boca Raton London New York

CRC Press is an imprint of the
Taylor & Francis Group, an **informa** business

Designed cover image: Shutterstock

First edition published 2024
by CRC Press
2385 Executive Center Drive, Suite 320, Boca Raton, FL 33431

and by CRC Press
4 Park Square, Milton Park, Abingdon, Oxon, OX14 4RN

CRC Press is an imprint of Taylor & Francis Group, LLC

ISBN: 978-1-032-46099-4 (hbk)
ISBN: 978-1-032-46419-0 (pbk)
ISBN: 978-1-003-38160-0 (ebk)

DOI: 10.1201/9781003381600

Typeset in Times
by Newgen Publishing UK

Contents

Chapter 7   Application of Industry 4.0, Digital Transformation, and Lean

*Jesús Vazquez-Hernandez, Rosario Martínez-García,
Fernando González-Aleu, Teresa Verduzco-Garza,
and Edgar M.A. Granda-Gutierrez*

Chapter 8   Application of Lean Six Sigma 4.0 in Seed Potato Value Chain

Gurraj Singh

Preface

This book provides significant knowledge about how to transform the conventional system into a digital one for operational excellence. It also provides significant coverage of the enablers or drivers of LSS 4.0, obstacles to LSS 4.0 implementation, tools, techniques, and the integration between LSS and I4.0. This book is a comprehensive attempt to sensitize readers to how industry can integrate LSS 4.0 into the existing product run and how LSS 4.0 implementation can be facilitated by comprehending different enablers and barriers. Additionally, the book describes key technologies and tool sets needed to execute a complete LSS 4.0 program. This book also covers the application of LSS 4.0 in different industrial sectors besides manufacturing. This book provides a complete guide, for all the stakeholders of manufacturing, to the different know-how, tools, methods, and key aspects of LSS 4.0. This book is a meticulous attempt by the book editors to encourage industry, managers, policymakers, governments, and human society to look after mother nature through a comprehensive understanding of the different hidden prospects associated with digitally enabled LSS.

About the Editors

Rajeev Rathi is Assistant Professor (Grade-1) in the Department of Mechanical Engineering, National Institute of Technology, Kurukshetra, India. His areas of research are production and operations management, Lean Six Sigma, decision making, Green Lean Six Sigma, circular economy, life cycle assessment, Industry 4.0, clean technologies measures and sustainable supply chain. He is adept in optimization and decision-making techniques like: Interpretive Structural Modelling, Preference Ranking Organization Method for Enrichment of Evaluations, Best Worst method, Decision-Making Trial and Evaluation Laboratory. He has more than ten years of teaching and research experience. He has actively engaged in handling industrial projects based on his expertise to achieve operational excellence. He has authored more than 70 research articles in highly reputed journals and conferences, ten book chapters and has four national and international patents granted. He is currently involved with special issues Q1 and Q2 journals as lead guest editor.

Jose Arturo Garza-Reyes is Professor of Operations Management and Head of the Centre for Supply Chain Improvement at the University of Derby, UK. He is actively involved in industrial projects where he combines his knowledge, expertise and industrial experience in operations management to help organizations achieve excellence in their internal functions and supply chains. He has also led and managed international research projects funded by the British Academy, British Council, European Commission and Mexico's National Council of Science and Technology (CONACYT). As a leading academic, he has published over 250 articles in leading scientific journals and international conferences and seven books. He is Associate Editor of the *International Journal of Operations and Production Management*, Associate Editor of the *Journal of Manufacturing Technology Management*, Editor of the *International Journal of Supply Chain and Operations Resilience* and Editor-in-Chief of the *International Journal of Industrial Engineering and Operations Management*. Areas of expertise and interest include general aspects of operations and manufacturing management, business excellence, quality improvement and performance measurement.

Mahender Singh Kaswan is Assistant Professor in the School of Mechanical Engineering, Lovely Professional University, Punjab, India. He holds a Ph.D. in Mechanical Engineering from Lovely Professional University, Punjab, M.Tech. in Industrial and Production Engineering from National Institute of Technology, Kurkshetra, India, and B.E. in Mechanical Engineering from the former Career Institute of Technology and Management, India (Manav Rachna International University). He has published several articles in leading international journals and conferences. His research interests include Lean Six Sigma, green manufacturing, life cycle assessment, circular economy, Industry 4.0, clean technology measures and sustainable supply chain.

Mahipal Singh is Associate Professor in the School of Mechanical Engineering, Lovely Professional University, Punjab, India. He holds a Ph.D. and M.Tech degree in Mechanical Engineering from Lovely Professional University, India. He holds a B.E. in Automobile Engineering from Maharishi Dayanand University, Rohtak, Haryana, India. He has published several articles in leading international journals and conferences. His research interests include Lean Six Sigma, green manufacturing, operational excellence, circular economy, Industry 4.0, clean technology measures and sustainable supply chain.

Contributors

Alexandre Acácio de Andrade
Postgraduate Program in Engineering and Management of Innovation, Federal University of ABC (UFABC), Santo André, São Paulo, Brazil

Abdelkabir Charkaoui
Faculty of Sciences and Technique, Hassan First University of Settat, Morocco

Anass Cherrafi
EST-Safi, Cadi Ayyad University, Marrakech-Safi, Morocco

Astha Dhall
Department of Forensic Science, School of Bioengineering and Biosciences, Lovely Professional University, Phagwara, Punjab, India

Júlio Francisco Blumetti Facó
Postgraduate Program in Engineering and Management of Innovation, Federal University of ABC (UFABC), Santo André, São Paulo, Brazil

Adriano Gomes de Freitas
Postgraduate Program in Engineering and Management of Innovation, Federal University of ABC (UFABC), Santo André, São Paulo, Brazil

Lab for Simulation and Modelling of Particulate Systems (SIMPAS), Monash University, Clayton, Melbourne, Australia

Jose Arturo Garza-Reyes
Centre for Supply Chain Improvement, The University of Derby, Derby, United Kingdom

Fernando González-Aleu
Department of Computer and Industrial Engineering, Universidad de Monterrey, San Pedro Garza García, NL, México

Rekha Goyat
School of Electronics and Electrical Engineering, Lovely Professional University, Punjab, India

Edgar M.A. Granda-Gutierrez
Department of Graduate School of Engineering and Technology, Universidad de Monterrey, San Pedro Garza García, NL, México

Mahender Singh Kaswan
School of Mechanical Engineering, Lovely Professional University, Phagwara, Punjab, India

Catarina de Andrade Lucizano
Postgraduate Program in Engineering and Management of Innovation, Federal University of ABC (UFABC), Santo André, São Paulo, Brazil

Rosario Martínez-García
WA Solutions, Monterrey, NL, México

Tejasvi Pandey
Department of Forensic Science, School of Bioengineering and Biosciences, Lovely Professional University, Phagwara, Punjab, India

Rajeev Rathi
Department of Mechanical Engineering, National Institute of Technology, Kurukshetra, Haryana, India

Geeta Sachdeva
Department of Humanities and Social Sciences, National Institute of Technology, Kurukshetra, Haryana, India

Ashutosh Samadhiya
Jindal Global Business School, OP Jindal Global University, Sonipat, Haryana, India

Bikram Jit Singh
Department of Mechanical Engineering, MM Engineering College, MMDU Mullana, Ambala, Haryana, India

Gurraj Singh
Department of Industrial and Production Engineering, Dr B.R. Ambedkar National Institute of Technology, Jalandhar, Punjab, India

Mahipal Singh
School of Mechanical Engineering, Lovely Professional University, Phagwara, Punjab, India

Dounia Skalli
Faculty of Sciences and Technique, Hassan First University of Settat, Morocco

Harsimran Singh Sodhi
Department of Mechanical Engineering, Chandigarh University, Gharuan, Punjab, India

Jesús Vazquez-Hernandez
Department of Communication Sciences, Universidad Autónoma de Nuevo León, San Nicolas de los Garza, México

Teresa Verduzco-Garza
Department of Computer and Industrial Engineering, Universidad de Monterrey, San Pedro Garza García, NL, México

Varun Vevek
Department of Forensic Science, School of Bioengineering and Biosciences, Lovely Professional University, Phagwara, Punjab, India

1 Overview of Lean Six Sigma 4.0 for Operational Excellence

Mahipal Singh, Rajeev Rathi, and Mahender Singh Kaswan

1.1 INTRODUCTION

When the first Industrial Revolution began, manufacturing changed from being a craft or cottage industry to mass production and then to lean manufacturing with leaner supply chains. Lean is a methodology for operational excellence that seeks to systematically eliminate waste through development of processes [1]. Further, the Lean approach was integrated with the Six Sigma approach, to minimize waste as well as process variations, and was termed as the first revolution of LSS and referred to as LSS 1.0 [2]. Snee and Hoerl introduced the "holistic" perspective of LSS in 2002, which transformed into the second revolution of LSS in 2018, which was termed LSS 2.0. Further development was done on LSS 2.0, termed as LSS 3.0, which revealed the improvement in the system that "can produce, sustain, and successfully integrate improvements in any environment, culture, and business" [3]. Additionally, LSS has become more technologically capable as manufacturing transforms into a more digitized environment with the development of Industry 4.0 (I4.0) technologies [4]. Consequently, the development of new I4.0 technologies like the Internet of Things (IoT), big data and data analytics, as well as augmented and virtual reality, may lead to the emergence of the fourth revolution in LSS, or LSS 4.0 [5]. To help the industrial system adopt the idea of "end-of-life," organizations are employing digital technology in closed loop supply chains to concentrate on the restorative and regenerative aspects [6]. Through the explicit usage of design models, product systems, and material design, these digital technologies ultimately aid in restoration, the removal of the use of hazardous goods, reuse, and the elimination of waste [7]. As a consequence, LSS 4.0 is now digitally enabled to address organizational objectives for the circular economy, by improving resource efficiency and environmental performance at various supply chain levels to achieve operational excellence. The term operational excellence is a term that is popular among most organizations and research areas at the present time. Operational excellence should be viewed as a framework and set of

DOI: 10.1201/9781003381600-1

tools that enable people within the business to operationalize change, according to a more recent invention.

1.2 KNOW-HOW ON LEAN SIX SIGMA 4.0

Lean Six Sigma is being widely used by industrial organizations to meet traditional quality characteristics. But to cope with modern challenges related to shorter product life cycle and production of customized high-quality sustainable products with minimum environmental damage, it is imperative to integrate digital technologies within the DMAIC (Define-Measure-Analyze-Improve-Control) methodology. LSS 4.0 is still in its infancy stage, and its successful implementation demands comprehension of the know-how of LSS 4.0. A comprehensive understanding of different enablers, success factors, barriers, tools, and technique knowledge is demanded to ensure the success of a novel approach.

1.3 ENABLERS OF LEAN SIX SIGMA 4.0

Enablers are elements that help move things forward or make it easier for someone to achieve something. LSS 4.0 is guided by several theoretical elements, including enablers, drivers, barriers, challenges, and critical failure factors. Therefore, these elements must be identified and understood for the most successful deployment of LSS 4.0. In this book, the critical success factors or enablers describe the vital components that must be taken into consideration by an organization on its way to LSS 4.0 and towards digital transformation.

1.4 BARRIERS TO LEAN SIX SIGMA 4.0

The implementation of LSS 4.0 is in its infancy stage across the world at present, and companies are facing vital issues in the adoption of LSS 4.0 in the running of the organization [8]. As there is not much information available about the barriers to the LSS 4.0 approach, industry professionals and practitioners face a constantly evolving difficulty in this area. Without information on the barriers/obstacles to the implementation of a new approach, it becomes really challenging to adopt the approach properly in the running system. This book presents the list of barriers to LSS 4.0 implementation.

1.5 ENABLING TECHNOLOGIES OF LEAN SIX SIGMA 4.0

LSS 4.0 as a continuous improvement approach with I4.0 is gaining much attention from experts in the field. Tissir et al. (2022) focus on identifying the LSS 4.0's conceptual underpinnings from its outward manifestation and categorizing them to guide further inquiry [9]. If corporations do not wait until that is done, they risk automating waste and increasing prices in the name of Industry 4.0. Without initially adopting LSS, businesses are free to deploy I4.0 solutions immediately. However, even though I4.0 technologies have the potential to minimize waste via automation, doing so may result in less-than-ideal solutions [10]. Artificial intelligence, machine learning, cloud computing, digital twin, and block chain are a few technologies that support implementation of LSS 4.0. Big data enabled with IoT devices can fetch huge data sets related to different process parameters for different products running on the machine that will further enhance the capability of the organization to optimize process parameters. This will ultimately help in the saving of energy and material for the organization which results in improved organizational environmental sustainability. Further, data grab through IoT devices will also help the organization to make different strategic decisions related to the process. Digital twin can also help the organization to imitate the process, and whether it will work in reality or not. This facility, associated with a digital twin, will further foster the organization's capability for effective utilization of scarce resources and ultimately contribute towards the environmental aspects of sustainability. AI and ML enable digital technologies to further foster organizational efforts to cope with the different challenges related to response time and quick decision making [11]. This book also pinpoints the key technologies of Industry 4.0 that finally help LSS to meet the modern challenges of industry related to customization of mass production products within environmental and social regulations.

1.6 APPLICATIONS OF LEAN SIX SIGMA 4.0 TO INDUSTRIAL SECTORS

The adoption of LSS 4.0 technologies in India as well as abroad has the potential to transform the country's manufacturing sector. The implementation of these technologies can improve the efficiency and productivity of the sector, reduce costs, and enable manufacturers to create more customized products and services [10]. However, adoption of the LSS 4.0 approach requires a skilled workforce, and manufacturers must invest in training and development programs to ensure that their employees have the necessary skills to operate and maintain these technologies. In the Indian context, the government has taken initiatives to promote the adoption of LSS 4.0 through launching the new campaigns like Make in India, Made in India, etc.

1.7 LESSON LEARNED AND INFERENCES

This book provides significant knowledge about how to transform the conventional system into a digital one for operational excellence. It also provides significant coverage of the enablers or drivers of LSS 4.0, obstacles to LSS 4.0 implementation, tools and techniques, and the integration between LSS and I4.0. This book is a comprehensive attempt to sensitize readers to how industry can integrate LSS 4.0 into the existing product run, and how LSS 4.0 implementation can be facilitated by comprehending different enablers and barriers. The book also describes the key technologies and tools needed to execute a complete LSS 4.0 program. This book also describes the application of LSS 4.0 in different industrial sectors besides manufacturing. This book provides a complete guide, for all the stakeholders of the manufacturing to the different know how, tools, methods, and key aspects of LSS 4.0. This book is a meticulous attempt by the authors to sensitize industry, managers, policy makers, governments, and all human society too look after mother nature through comprehensive understanding of the different hidden prospects associated with digitally enabled LSS.

REFERENCES

[1] O. McDermott, J. Antony, M. Sony, and S. Daly, "Barriers and Enablers for Continuous Improvement Methodologies within the Irish Pharmaceutical Industry", *Processes*, vol. 10, no. 1, 2022. doi: 10.3390/pr10010073.

[2] M. L. George, "Lean six sigma: combining six sigma quality with lean speed", McGraw-Hill, pp. 17–22, 2002.

[3] R. D. Snee and R. Hoerl, *Leading Holistic Improvement with Lean Six Sigma 2.0*. FT Press, 2018.

[4] A. Calabrese, M. Dora, N. Levialdi Ghiron, and L. Tiburzi, "Industry's 4.0 transformation process: how to start, where to aim, what to be aware of", *Prod. Plan. Control*, vol. 33, no. 5, pp. 492–512, 2022. doi:10.1080/09537287.2020.1830315.

[5] G. Arcidiacono and A. Pieroni, "The revolution Lean Six Sigma 4.0", *Int. J. Adv. Sci. Eng. Inf. Technol.*, vol. 8, no. 1, pp. 141–149, 2018. doi:10.18517/ijaseit.8.1.4593.

[6] A. B. Lopes de Sousa Jabbour, C. J. C. Jabbour, M. Godinho Filho, and D. Roubaud, "Industry 4.0 and the circular economy: a proposed research agenda and original roadmap for sustainable operations", *Ann. Oper. Res.*, vol. 270, no. 1–2, pp. 273–286, 2018. doi:10.1007/s10479-018-2772-8.

[7] S. Rajput and S. P. Singh, "Connecting circular economy and industry 4.0", *Int. J. Inf. Manage.*, vol. 49, pp. 98–113, 2019. doi:10.1016/j.ijinfomgt.2019.03.002.

[8] I. P. Vlachos, R. M. Pascazzi, G. Zobolas, P. Repoussis, and M. Giannakis, "Lean manufacturing systems in the area of Industry 4.0: a lean automation plan of AGVs/IoT integration", *Prod. Plan. Control*, vol. 34, no. 4, 2021. doi:10.1080/09537287.2021.1917720.

[9] S. Tissir, A. Cherrafi, A. Chiarini, S. Elfezazi, and S. Bag, "Lean Six Sigma and industry 4.0 combination: Scoping review and perspectives", Total Quality Management & Business Excellence, 1–30, 2022.

[10] J. E. Sordan, P. C. Oprime, M. L. Pimenta, S. L. da Silva, and M. O. A. González, "Contact points between Lean Six Sigma and Industry 4.0: a systematic review and conceptual framework", *International Journal of Quality and Reliability Management*, vol. 39, no. 9. pp. 2155–2183, 2022. doi:10.1108/IJQRM-12-2020-0396.

[11] A. Chiarini and M. Kumar, "Lean Six Sigma and Industry 4.0 integration for Operational Excellence: evidence from Italian manufacturing companies", *Prod. Plan. Control*, vol. 32, no, 13 2020. doi:10.1080/09537287.2020.1784485.

2 Connection between Lean Six Sigma (LSS) and Industrial IoT for Operational Excellence

Rekha Goyat, Mahipal Singh, and Rajeev Rathi

2.1 INTRODUCTION

Due to financial challenges, industries are focusing on various technologies to improve their operations and performance [1]. In this milieu, it has been said that industries can accelerate their performance and operations by adopting the various skills under Industry 4.0 (I4.0) [2]. Also, Continuous Improvement (CI) schemes are adopted within organizations to enhance their operation and processes with the least resources. Various CI practices such as Kaizen, Lean, Six Sigma (SS), and Lean Six Sigma (LSS) are increasing the capability to produce top class quality products. Adopting LSS within industry is another way to achieve excellence and improve operations. LSS is a well-known tactic to maximize productivity, the quality of products, speed, and customer satisfaction by reducing waste, defects, and overall cost [3]. Due to this, the amalgam of Industry 4.0 and LSS has become an exciting area of research for the researchers and academia. Manufacturing industries are focusing on the fourth industrial revolution. The impact of a combined approach could result in a big revolution. The integration is beneficial to get maximum results by utilizing available resources optimally. Industry 4.0 was first introduced by fair technology in Hannover with the aim to boost productivity and ability to compete in various industries globally [4]. Integration of digital perspectives with manufacturing enables the industry by creating intelligent and independent processes within the system, and allowing direct communication of machines to each other by using various technologies such as Internet of Things (IoTs), artificial intelligence, big data, machine learning, blockchain, cloud-computing, and cyber-physical systems [5][6]. Basically, automation within the processes completely renovates the system by enabling the mass customization of products, replacement of layouts, increased production speed of items and communication among machinery for controlling processes. A report generated in partnership with the World Economic Forum states that by introducing Industry 4.0 within processes, it is possible to generate an added value of 3.7 trillion dollars in revenue for 2025. In other words, Industry 4.0 is totally revolving around the Internet, production flexibility, and virtualization of the systems. However, there is no perfect harmony or perfect definition of Industry 4.0.

DOI: 10.1201/9781003381600-2

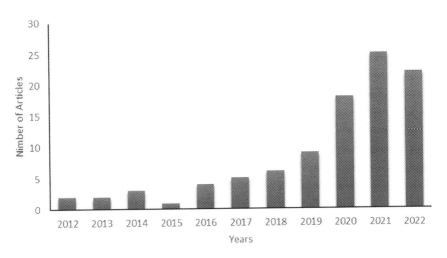

FIGURE 2.1 Article count published per year.

On the other side, LSS is a continuous improvement technique which minimizes wastage and process variations and results in higher productivity [7]. It first came into existence in the early 2000's through the amalgamation of two popular practices: Lean manufacturing (L), and Six Sigma (SS) [8]. While the focus of both methods is process improvement, they work with different criteria. The main motive of L is to eliminate wastage and non-value-added processes, while SS aims to reduce process variability to improve the quality of products.

Therefore, industries become more competitive and responsive, as per market demand, due to the integration of Industry 4.0 with LSS. Also, Industry 4.0 combined with Lean and LSS has become the most popular topic among practitioners, researchers, and academics. The relationship between Industry 4.0 and Continuous Improvement approaches is well investigated in the literature as shown in Figure 2.1, but most of the studies have discussed the association between Industry 4.0 and Lean. Therefore, more studies addressing the amalgamation of Industry 4.0 with LSS are required to enable better implementation within organizations.

2.2   THEORETICAL BACKGROUND

2.2.1   INDUSTRIAL IoT

Recently, the Industrial IoT (IIoT) has proved its uniqueness by using machinery, cloud computing, Artificial Intelligence, and people together to improve the productivity, quality and performance of industrial processes [9]. IIoT has been introduced in various industries, including manufacturing, health, transportation, and utilities, as an operational technology. The IoT was first established in 1999 by Kevin Ashton and applied to connected devices in domestic, business, consumer and industry scenarios [10]. However, it became more popular in summer 2010 and an organization named

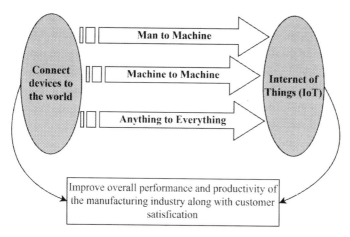

FIGURE 2.2   Industrial Internet of Things.

Gartner included this emerging phenomenon for market research. After that, the theme of the conference named 'LeWeb', held in Europe, was IoT and it gained more popularity across the world [11]. In IIoT, sensor nodes are deployed in the environment, gathering data from various pieces of equipment to measure them and control them accordingly. In other words, it is the interconnection of devices and their data using internet connection, without the intervention of human-to-human association. Industrial IoT is playing a crucial role over the last decade without linking to LSS [12]. The association between data, and its processing, with IIoT provides more evident and significant results [13]. With the amalgamation of vertical, horizontal, and end-to-end association, manufacturing sectors are expected to change and augment overall productivity and quality by up to 55%, and profits by up to 15% [14][15]. Figure 2.2 reveals that physical machinery, and the associated processes, will have embedded smart sensors, actuators, and automatic software that will enable fast processing and communication of data in real time.

IIoT is a group of devices connected to different machines equipped with sensors and actuators for sensing different physical activity happening in the environment with minimal human interference. Each sensor node in a virtual environment continuously transmits a large volume data about itself and surrounding nodes.

2.2.2 Lean Six Sigma

In today's world, industries are facing various challenges to meet customers' requirements with high quality products. Customers need for higher quality products at lower cost has compelled industries to think about their profits. Each organization wants to earn more and for that they want to adopt quality schemes that enhance overall productivity and quality, whilst reducing overall manufacturing cost. The overall profits of the industry depend on the capability to produce the highest quality products at low-cost [16]. To produce high quality products with least-cost, Lean Six Sigma (LSS) is one popular approach based on Lean which aims to minimize waste, and Six Sigma which aims to reduce process variations [16]. LSS as a continuous improvement approach has been implemented in numerous manufacturing and service organizations across the globe [17]. This approach is able to reduce the non-value-added activities and variability in the production line to obtain superior quality at optimal cost. The LSS approach is primarily executed through the DMAIC (Define-Measure-Analyze-Improve-Control) methodology in most of the projects. The literature also reveals that the DMAIC approach has been implemented for continuous improvement in process industries [18], the manufacturing industry [19], the printing industry [20], healthcare industries [21], etc. In recent years, practitioners have increasingly focused on integrating advanced technologies like Industry 4.0 with continuous improvement methods and redesigning their processes to improve operational performance [22]. However, the literature still lacks information on the integration of LSS and the Industrial IoT due to numerous challenges [23]. In this context, this chapter aims to propose a possible integration among LSS and the Industrial IoT with a list of possible enablers and barriers.

2.2.3 INDUSTRIAL IoT AND LSS INTEGRATION

Due to rapid developments in smart devices and internet-based devices, sensor nodes can perceive the irregularity and malfunctioning of a machine in the production line in real time and can alert the controlling authority. Due to advances in technology, business and industries have become more efficient over the last decades. An organization that doesn't adopt these advances will face a lot of challenges and difficulties in competing. Still, the adoption of technology is useless in the absence of coordinated or smart human processes like LSS. Therefore, the integration of Industrial IoT with LSS enables the system to create effective and efficient conditions that support each other during their implementation to get optimal results. The IIoT provides a widespread data set that can be extracted in order to enable smooth quality improvement procedures to take place during processing. The extracted data can be utilized by LSS to minimize waste reduction and improve operational efficiency. The main objectives of IIoT and LSS are the same which facilitates their use in a cooperative plan in order to get more efficiency within industry. The LSS approach eliminates defects and waste to improve the processes, but the IIoT supports by achieving deeper insights within the processes.

In other words, LSS that is executed based on collected information and information benefits from the visibility of the working environment of the IIoT. The IIoT gathers the real-time data from the working environment and the same data can be utilized by the LSS approach to convert data into actionable business intelligence. The gathered data enables the controlling authority to take corrective measures to improve the performance and productivity within the organization. Also, the non-added value happenings can be identified accurately which ensures superior production quality. Table 2.1 describes the related work in the field of Industry 4.0 and

TABLE 2.1
Summary Description of Relevant Work

Technique	Description	Year of Publication	Author
Integration of Lean and Industry 4.0	Industry 4.0 and Lean management introduces closely related terms and conditions to practical implementation to improve the production quality within the industry.	2022	[24]
Lean 4.0 tool with digital technology	In this paper, the possible benefits, challenges, implementation protocols and trends of integrated Lean with Industry 4.0 have been discussed. Also, the characterization of the Lean 4.0 tool with digital advancements has been explored.	2022	[25]
Integration of LSS with Industry 4.0 in Italian company	Integration of LSS with Industry 4.0 is applied at an Italian manufacturing company to achieve effective outcome. The embedded approaches needed reinvented mapping tools for horizontal and vertical end-to-end integration.	2021	[26]
Continuous Improvement (CI) practices with Industry 4.0	The amalgam of Continuous Improvement (CI) practices with Industry 4.0 is contributing in a new way to gain more effective and strategic benefits.	2020	[27]
LSS with Industry 4.0	The relationship between LSS and Industry 4.0 has been investigated in this Moroccan context-based survey. This survey conveys that both the approaches are more compatible and synergetic during their implementation.	2019	[2]
Digitization and automation in Industry 4.0 for Lean	Industry 4.0 is trending currently for automation and digitization within the manufacturing sector and is mainly utilizing enabling technologies IoT, big data, blockchain, and cloud computing.	2018	[28]
Lean and simulation-based optimization with Industry 4.0	Two approaches Lean and simulation-based optimization are combined with Industry 4.0 for optimal efficiency and wasted reduction.	2017	[29]
Industry 4.0 implementation in Lean manufacturing	The paper provides crucial insights to manufacturers for adopting Industry 4.0 within the organization with possible benefits, challenges, and outcomes.	2016	[30]
Lean automation with Industry 4.0	In this paper, major corner stones and connections of Industry 4.0 have been discussed to the well-known approach Lean.	2015	[31]

Continuous Improvement approaches, but the IIoT as an enabler of CI is not explored much in the literature.

In other words, we can say that the integration of IIoT with LSS has received moderate attention from researchers and academia. Furthermore, the possible challenges, barriers, enablers, and motivation around Industry 4.0 and LSS have been investigated through a literature review. After identifying the research gaps, we propose a systematic research design for evaluating the literature to find out the list of barriers and enablers of IIoT and LSS.

2.3   RESEARCH METHODOLOGY

In this section, a Systematic Literature Review (SLR) scheme is used to identify the barriers and enablers of IIoT and LSS integration. This scheme ensures a high level of precision, transparency, intelligence, and objectivity during the literature review process. The SLR was basically completed in three steps i.e. 1) Planning the review, 2) Conducting the review with inclusion and exclusion criteria and 3) finally Discussion of collected information in relation to the future research agenda as shown in Figure 2.3.

To attain the research objectives, the research papers were downloaded from various databases such as ScienceDirect, Scopus, Taylor & Francis, ResearchGate, Google Scholar, Pro-Quest and Informs etc. Various keywords such as Industrial IoT, Industry 4.0, integrate Industry 4.0 with Lean, Six Sigma, and LSS tools and techniques were used to identify the relevant research papers. After the identification process, a worksheet was arranged to exclude the duplicate and non-relevant

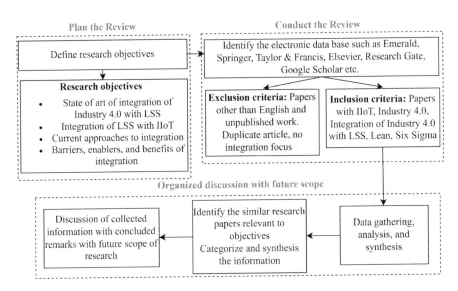

FIGURE 2.3   Structure of systematic literature review.

TABLE 2.2
Inclusion and Exclusion Criteria for Identifying Relevant Research Papers

Inclusion Criteria	Exclusion Criteria
• Papers with IIoT and Industry 4.0 • Integration of Industry 4.0 with LSS, Lean, Six Sigma • Tools and techniques of LSS, Lean, Lean Six Sigma • Barriers and enablers of integration	• Papers others than English and unpublished work • Duplicate article • No integration focus • No full-text available • Non-peer-reviewed papers

papers. The papers were carefully examined to certify that papers were only selected based on the inclusion and exclusion criteria. A detailed analysis of the title, abstract, keywords, and introduction was executed to gather the more accurate and relevant papers. The various inclusion criteria and exclusion criteria used for analysis are discussed in Table 2.2.

After that, analysis and compilation of the articles was carried out to explore the barriers, enablers, and benefits of integrating IIoT with LSS. At the end, the conclusions of the study were discussed in relation to the future research direction.

2.4 RESULTS AND DISCUSSION

2.4.1 Enablers of Integrated IIoT with LSS

In this section, the enablers of integrating IIoT and LSS are discussed. Enablers represent the factors or conditions that ensure the possibility of integration. The list of enablers that has been extracted after a detailed systematic literature review is discussed in Table 2.3.

Enabling technologies and tools that ensure the smooth operation of integrated IIoT with LSS are discussed in Table 2.4. IIoT deliberately uses various enabling technologies for controlling and automating processes and these technologies are WSNs, BLE, RFID, Wi-Fi, WMSNs etc.

2.4.2 Barriers of Integrated IIoT with LSS

Although industries are more aware and informed, the role of digitization and automation in adding value, as well as the numerous benefits to manufacturing sectors, are not fully recognized [63]. Reaping the potential and benefits of IIoT in industry is not achievable without understanding a suitable execution procedure, which needs a thorough preadaptation analysis, including identification of all possible barriers and challenges [64][65]. The list of barriers to adoption of IIoT with LSS within an organization are discussed in Table 2.5.

TABLE 2.3
Enablers of Integrated IIoT with LSS

Notation	Enablers	Descriptions	Literature Support
EN 1	Involvement of staff, workers, and teams	Involvement of staff, workers, and teams helps to attain company goal in an energetic manner and encourage management to adopt integrated scheme.	[32]
EN 2	Management support and commitment	Management support and commitment affects the teammates positively and encourages company willingness to adopt IIoT with LSS.	[33]
EN 3	Availability of staff	Involvement of staff affects the outcome of the industry and gives the commitment pf management as to why integration is necessary. The availability and support of staff boost the working empowerment and partnership in a problem-solving environment.	[34]
EN 4	Training of staff	Training of staff may facilitate the understanding of IIoT and LSS in depth. Also, it encourages the management to recognize the profits of integrated IIoT and LSS adoption.	[35]
EN 5	Fund for seed and infrastructure	Fund for infrastructure can facilitate the adoption of an integrated scheme by providing the financial support to the industry for technological acquisition.	[32]
EN 6	Implementation of integrated strategy with business strategy	The implementation of an integrated scheme within the organization is a critical factor that affects the outcomes positively by approaching business strategies within it.	[36]
EN 7	Knowledge of IIoT and LSS	The detailed information and knowledge about the benefits of IIoT and LSS play a crucial role that inspires management for adoption.	[37]
EN 8	High company maturity in the use of IIoT and LSS	Maturity of industry about IIoT and LSS can facilitate in favor of integration adoption.	[38]

EN 9	Availability of implementation resources for integration	The availability of resources affects the implementation criteria within the firm. Availability of resources makes it easy for industry to adopt integration.	[39]
EN 11	Higher standard of operation	Higher standards of operation enable smooth processing within the process and improve productivity of the industry.	[40]
EN 12	Regulation of policies	Regulation of policies encourages the management to move from a traditional production model to a systematic and automatic model through integrated IIoT and LSS.	[41]
EN 13	Protection and security of information	The innovative and sustainable technology gives protection to the confidential information and also ensures the security within the firm.	[42] [40][38]
EN 15	Real-time data availability	The IIoT technology enables the real-time monitoring of data by deploying sensor nodes.	[43]
EN 16	Transparency of systems	IIoT gives more transparency to the system which is the most crucial feature of digitalization.	[44]
EN 17	Flexibility of digitization	Digital technologies are more flexible and easier to activate without human intervene.	[45][38]

TABLE 2.4
Enabling Technologies and Tools for IIoT and LSS

Enabling Technologies for IIoT	Literature Support	Enabling Technologies for LSS	Literature Support
Wireless Sensor Networks (WSNs)	[46]	Just-in-Time	[47]
Wireless Multimedia Sensor Networks (WMSNs)	[48]	Kanban	[49]
Bluetooth Low Energy (BLE)	[10]	Kaizen	[50]
Radio-Frequency Identification (RFID)	[51]	5S	[52]
Digital automation/Sensors	[53]	Poka-Yoke	[54]
Zigbee	[55]	Total Quality Management	[56]
Wi-Fi	[57]	FMEA	[58]
Machine Learning	[59]	VSM	[60]
Artificial intelligence	[61]	SPC	[62]

TABLE 2.5
Barriers of Integrated IIoT with LSS

Notation	Barriers	Description	Literature Support
BA 1	Resistance to change	Management is not ready to admit the changes that are needed for implementing integrated IIoT and LSS.	[66]
BA 2	Lack of awareness	The staff and management have a lack of awareness about advanced techniques and tools.	[67][68]
BA 3	Insufficient management support	Management inadvertently not supporting and not allowing workers and employs to develop and build their skills and failing to provide support to team.	[69][70]
BA 4	Lack of organizational communication	Lack of organizational communication is a failure that results in incongruity between teammates and management.	[71][70]
BA 5	Concerns with collaboration and network formation	Concerns with collaboration is a major barrier which affects communication and hinder the benefits of network formation.	[66] [72]
BA 6	Lack of competence nature and staff complications	Lack of competence nature and staff complications hinders the advantages of adoption of integrated IIoT and LSS within firm.	[73][68]

TABLE 2.5 (*Continued*)
Barriers of Integrated IIoT with LSS

Notation	Barriers	Description	Literature Support
BA 7	Short-term vision and mission of industry	A mission and vision describe the objective of industry. Short-term vision is a failure that results in inconsistency.	[69][74]
BA 8	Lack of infrastructure	Poor infrastructure in industry hinders the benefits of integrated IIoT and LSS.	[75]
BA 9	Tough learning process	The integration procedure is based on advanced techniques and tools. Staff must have knowledge and experience which is hard to learn.	[70][69]
BA 10	Data storing, security and complexity issues	Storing and security of data relevant to industry are issues of concern and can be considered as a weakness to implementing IIoT with LSS.	[71]
BA 11	Unclear protocols and standards	Management is somehow unwilling due to unclear protocols and standards in the integrated scheme.	[76]
BA 12	Technological incompatibility	Technological incompatibility and maturity level are responsible for not adopting the integrated scheme within the organization.	[77]
BA 13	Workforce uncertainty	Management is not ready to adopt integrated scheme due to uncertainty of workforce and regular fluctuations in processes.	[78]
BA 14	Staff with least experience	Staff experience and infancy about an integrated scheme hinders the advantages of adoption of IIoT with LSS within the organization.	[76]
BA 15	Uncertain profits	The uncertainty of profits and benefits to industry creates resistance in the management for IIoT and LSS adoption.	[75]
BA 16	Incompatibility among integrated schemes	Different schemes cannot communicate easily and the absence of standardized protocols hinder the profits of an integrated scheme.	[73][79]

2.5 CONCLUSION

Industry 4.0 simply represents the era of digitization and automation, which endorses independent and autonomous processes to achieve the rapid growth of industry. The processes and operations in industry are performed in an automatic manner without human intervention. Industrial IoT is the most important technology and acts as the backbone of Industry 4.0. Industrial IoT makes more effective and fact-based decisions when deployed with LSS. Integrating IIoT with LSS can provide valuable insight into the data collected from the organization, and results in quality and productivity improvement. Therefore, it is essential to understand the potential benefits, enablers, barriers, and tools faced during the deployment of IIoT with LSS. The present chapter is restricted to the searching criteria, literature search, inclusion and exclusion criteria which can be replicated in opportunities for future work.

REFERENCES

[1] A. Cherrafi, S. Elfezazi, A. Chiarini, A. Mokhlis, and K. Benhida, "The integration of lean manufacturing, Six Sigma and sustainability: A literature review and future research directions for developing a specific model", *J. Clean. Prod.*, vol. 139, pp. 828–846, 2016. doi: 10.1016/j.jclepro.2016.08.101.

[2] C. Anass, B. Amine, E. H. Ibtissam, I. Bouhaddou, and S. Elfezazi, "Industry 4.0 and Lean Six Sigma: Results from a Pilot Study", in *Lecture Notes in Mechanical Engineering*, pp. 613–619, 2021. doi: 10.1007/978-3-030-62199-5_54.

[3] K. F. Barcia, L. Garcia-Castro, and J. Abad-Moran, "Lean Six Sigma Impact Analysis on Sustainability Using Partial Least Squares Structural Equation Modeling (PLS-SEM): A Literature Review", *Sustainability (Switzerland)*, vol. 14, no. 5. 2022. doi: 10.3390/su14053051.

[4] A. Firu, A. Tapirdea, O. Chivu, A. I. Feier, and G. Draghici, "The competences required by the new technologies in industry 4.0 and the development of employees' skills", *ACTA Tech. NAPOCENSIS Ser. Math. Mech. Eng.*, vol. 64, no. 1, p. 109–116 WE–Emerging Sources Citation Index (ESC), 2021.

[5] K. Shafique, B. A. Khawaja, F. Sabir, S. Qazi, and M. Mustaqim, "Internet of things (IoT) for next-generation smart systems: A review of current challenges, future trends and prospects for emerging 5G-IoT Scenarios", *IEEE Access*, vol. 8. pp. 23022–23040, 2020. doi: 10.1109/ACCESS.2020.2970118.

[6] V. Filimonau and E. Naumova, "The blockchain technology and the scope of its application in hospitality operations", *Int. J. Hosp. Manag.*, 2019. doi: 10.1016/j.ijhm.2019.102383.

[7] M. Singh, R. Rathi, and J. A. Garza-Reyes, "Analysis and prioritization of Lean Six Sigma enablers with environmental facets using best worst method: A case of Indian MSMEs", *J. Clean. Prod.*, 2021. doi: 10.1016/j.jclepro.2020.123592.

[8] G. Yadav and T. N. Desai, "Analyzing Lean Six Sigma enablers: A hybrid ISM-fuzzy MICMAC approach", *TQM J.*, vol. 29, no. 3, pp. 488–510, 2017. doi: 10.1108/TQM-04-2016-0041.

[9] E. Manavalan and K. Jayakrishna, "A review of Internet of Things (IoT) embedded sustainable supply chain for industry 4.0 requirements", *Comput. Ind. Eng.*, vol. 127, pp. 925–953, 2019. doi: 10.1016/j.cie.2018.11.030.

[10] V. Chang and C. Martin, "An industrial IoT sensor system for high-temperature measurement", *Comput. Electr. Eng.*, vol. 95, 2021. doi: 10.1016/j.compeleceng.2021.107439.

[11] P. R. Newswire, "Global Industrial IoT Market: Research report 2015–2019", *Lon-Reportbuyer*, 2015.

[12] Q. Chi, H. Yan, C. Zhang, Z. Pang, and L. Da Xu, "A reconfigurable smart sensor interface for industrial WSN in IoT environment", *IEEE Trans. Ind. Informatics*, vol. 10, no. 2, pp. 1417–1425, 2014. doi: 10.1109/TII.2014.2306798.

[13] A. M. A. Zamil, A. Al Adwan, and T. G. Vasista, "Enhancing customer loyalty with market basket analysis using innovative methods: a python implementation approach", *Int. J. Innov. Creat. Chang.*, vol. 14, no. 2, pp. 1351–1368, 2020.

[14] A. Raj, G. Dwivedi, A. Sharma, A. B. Lopes de Sousa Jabbour, and S. Rajak, "Barriers to the adoption of industry 4.0 technologies in the manufacturing sector: An inter-country comparative perspective", *Int. J. Prod. Econ.*, vol. 224, 2020. doi: 10.1016/j.ijpe.2019.107546.

[15] P.-L. Caylar, "Digital in industry: From buzzword to value creation", *McKinsey Digit.*, no. August, pp. 1–9, 2016.

[16] M. Singh and R. Rathi, "A structured review of Lean Six Sigma in various industrial sectors", *International Journal of Lean Six Sigma*. 2019. doi: 10.1108/IJLSS-03-2018-0018.

[17] M. Singh, R. Rathi, J. Antony, and J. A. Garza-Reyes, "A toolset for complex decision-making in analyze phase of Lean Six Sigma project: a case validation", 2022. doi: 10.1108/IJLSS-11-2020-0200.

[18] S. Bhat, E. V. Gijo, and N. A. Jnanesh, "Application of Lean Six Sigma methodology in the registration process of a hospital", *Int. J. Product. Perform. Manag.*, vol. 63, no. 5, pp. 613–643, 2014. doi: 10.1108/IJPPM-11-2013-0191.

[19] M. K. Hassan, "Applying Lean Six Sigma for Waste Reduction in a Manufacturing Environment", *Am. J. Ind. Eng.*, vol. 1, no. 2, pp. 28–35, 2013. doi: 10.12691/AJIE-1-2-4.

[20] N. Roth and M. Franchetti, "Process improvement for printing operations through the DMAIC lean six sigma approach: A case study from northwest Ohio, USA", *Int. J. Lean Six Sigma*, vol. 1, no. 2, pp. 119–133, 2010. doi: 10.1108/20401461011049502.

[21] P. Kumar, D. Singh, and J. Bhamu, "Development and validation of DMAIC based framework for process improvement: a case study of Indian manufacturing organization", *Int. J. Qual. Reliab. Manag.*, vol. 38, no. 9, pp. 1964–1991, 2021. doi: 10.1108/IJQRM-10-2020-0332.

[22] A. Chiarini and M. Kumar, "Lean Six Sigma and Industry 4.0 integration for Operational Excellence: evidence from Italian manufacturing companies", *Prod. Plan. Control*, 2020. doi: 10.1080/09537287.2020.1784485.

[23] J. E. Sordan, P. C. Oprime, M. L. Pimenta, S. L. da Silva, and M. O. A. González, "Contact points between Lean Six Sigma and Industry 4.0: a systematic review and conceptual framework", *International Journal of Quality and Reliability Management*, vol. 39, no. 9. pp. 2155–2183, 2022. doi: 10.1108/IJQRM-12-2020-0396.

[24] T. Komkowski, J. Antony, J. A. Garza-Reyes, G. L. Tortorella, and T. Pongboonchai-Empl, "The integration of Industry 4.0 and Lean Management: a systematic review and constituting elements perspective", *Total Qual. Manag. Bus. Excell.*, pp. 1–18, 2022. doi: 10.1080/14783363.2022.2141107.

[25] A. H. G. Rossi *et al.*, "Lean Tools in the Context of Industry 4.0: Literature Review, Implementation and Trends", *Sustain.*, vol. 14, no. 19, 2022. doi: 10.3390/su141912295.

[26] A. Chiarini and M. Kumar, "Lean Six Sigma and Industry 4.0 integration for Operational Excellence: evidence from Italian manufacturing companies", *Prod. Plan. Control*, vol. 32, no. 13, pp. 1084–1101, 2021. doi: 10.1080/09537287.2020.1784485.

[27] S. Vinodh, J. Antony, R. Agrawal, and J. A. Douglas, "Integration of continuous improvement strategies with Industry 4.0: a systematic review and agenda for further research", *TQM Journal*, vol. 33, no. 2. pp. 441–472, 2021. doi: 10.1108/TQM-07-2020-0157.

[28] G. Xu, M. Li, C. H. Chen, and Y. Wei, "Cloud asset-enabled integrated IoT platform for lean prefabricated construction", *Autom. Constr.*, vol. 93, pp. 123–134, 2018. doi: 10.1016/j.autcon.2018.05.012.

[29] E. R. Zuniga, M. U. Moris, and A. Syberfeldt, "Integrating simulation-based optimization, lean, and the concepts of industry 4.0", in *Proceedings – Winter Simulation Conference*, pp. 3828–3839, 2017. doi: 10.1109/WSC.2017.8248094.

[30] A. Sanders, C. Elangeswaran, and J. Wulfsberg, "Industry 4.0 implies lean manufacturing: Research activities in industry 4.0 function as enablers for lean manufacturing", *J. Ind. Eng. Manag.*, vol. 9, no. 3, pp. 811–833, 2016. doi: 10.3926/jiem.1940.

[31] D. Kolberg and D. Zühlke, "Lean Automation enabled by Industry 4.0 Technologies", in *IFAC-PapersOnLine*, vol. 28, no. 3, pp. 1870–1875, 2015. doi: 10.1016/j.ifacol.2015.06.359.

[32] G. Tortorella, R. Miorando, R. Caiado, D. Nascimento, and A. Portioli Staudacher, "The mediating effect of employees' involvement on the relationship between Industry 4.0 and operational performance improvement", *Total Qual. Manag. Bus. Excell.*, vol. 32, no. 1–2, pp. 119–133, 2021. doi: 10.1080/14783363.2018.1532789.

[33] T. Ilangakoon, S. Weerabahu, and R. Wickramarachchi, "Combining Industry 4.0 with Lean Healthcare to Optimize Operational Performance of Sri Lankan Healthcare Industry", 2019. doi: 10.1109/POMS.2018.8629460.

[34] H. Saabye, T. B. Kristensen, and B. V. Wæhrens, "Real-time data utilization barriers to improving production performance: An in-depth case study linking lean management and industry 4.0 from a learning organization perspective", *Sustain.*, vol. 12, no. 21, pp. 1–21, 2020. doi: 10.3390/su12218757.

[35] J. Abbas, "Impact of total quality management on corporate sustainability through the mediating effect of knowledge management", *J. Clean. Prod.*, vol. 244, 2020. doi: 10.1016/j.jclepro.2019.118806.

[36] S. Shamim, S. Cang, H. Yu, and Y. Li, "Examining the feasibilities of Industry 4.0 for the hospitality sector with the lens of management practice", *Energies*, vol. 10, no. 4, 2017. doi: 10.3390/en10040499.

[37] B. Busto Parra, P. Pando Cerra, and P. I. Álvarez Peñín, "Combining ERP, Lean Philosophy and ICT: An Industry 4.0 Approach in an SME in the Manufacturing Sector in Spain", *EMJ – Eng. Manag. J.*, 2021. doi: 10.1080/10429247.2021. 2000829.

[38] M. Ghobakhloo and M. Fathi, "Corporate survival in Industry 4.0 era: the enabling role of lean-digitized manufacturing", *J. Manuf. Technol. Manag.*, vol. 31, no. 1, pp. 1–30, 2020. doi: 10.1108/JMTM-11-2018-0417.

[39] M. Rossini, F. Costa, G. L. Tortorella, A. Valvo, and A. Portioli-Staudacher, "Lean Production and Industry 4.0 integration: how Lean Automation is emerging in manufacturing industry", *Int. J. Prod. Res.*, vol. 60, no. 21, pp. 6430–6450, 2022. doi: 10.1080/00207543.2021.1992031.

[40] G. L. Tortorella, M. Rossini, F. Costa, A. Portioli Staudacher, and R. Sawhney, "A comparison on Industry 4.0 and Lean Production between manufacturers from emerging and developed economies", *Total Qual. Manag. Bus. Excell.*, vol. 32, no. 11–12, pp. 1249–1270, 2021. doi: 10.1080/14783363.2019.1696184.

[41] F. D. Cifone, K. Hoberg, M. Holweg, and A. P. Staudacher, "'Lean 4.0': How can digital technologies support lean practices?", *Int. J. Prod. Econ.*, vol. 241, 2021. doi: 10.1016/j.ijpe.2021.108258.

[42] G. Tortorella, R. Sawhney, D. Jurburg, I. C. de Paula, D. Tlapa, and M. Thurer, "Towards the proposition of a Lean Automation framework: Integrating Industry 4.0 into Lean Production", *J. Manuf. Technol. Manag.*, vol. 32, no. 3, pp. 593–620, 2021. doi: 10.1108/JMTM-01-2019-0032.

[43] G. Tortorella, R. Miorando, and A. F. Mac Cawley, "The moderating effect of Industry 4.0 on the relationship between lean supply chain management and performance improvement", *Supply Chain Manag.*, vol. 24, no. 2, pp. 301–314, 2019. doi: 10.1108/SCM-01-2018-0041.

[44] S. Mittal, M. A. Khan, D. Romero, and T. Wuest, "Smart manufacturing: Characteristics, technologies and enabling factors", *Proc. Inst. Mech. Eng. Part B J. Eng. Manuf.*, vol. 233, no. 5, pp. 1342–1361, 2019. doi: 10.1177/0954405417736547.

[45] J. Ma, Q. Wang, and Z. Zhao, "SLAE–CPS: Smart lean automation engine enabled by cyber-physical systems technologies", *Sensors (Switzerland)*, vol. 17, no. 7, 2017. doi: 10.3390/s17071500.

[46] R. Goyat, G. Kumar, M. Alazab, R. Saha, R. Thomas, and M. K. Rai, "A secure localization scheme based on trust assessment for WSNs using blockchain technology", *Futur. Gener. Comput. Syst.*, vol. 125, pp. 221–231, 2021. doi: 10.1016/j.future.2021.06.039.

[47] M. S. Kaswan, R. Rathi, and M. Singh, "Just in time elements extraction and prioritization for health care unit using decision making approach", *Int. J. Qual. Reliab. Manag.*, 2019. doi: 10.1108/IJQRM-08-2018-0208.

[48] F. Bouakkaz, W. Ali, and M. Derdour, "Forest fire detection using wireless multimedia sensor networks and image compression", *Instrum. Mes. Metrol.*, vol. 20, no. 1, pp. 57–63, 2021. doi: 10.18280/I2M.200108.

[49] R. Sundar, A. N. Balaji, and R. M. Satheesh Kumar, "A review on lean manufacturing implementation techniques", *Procedia Eng.*, vol. 97, pp. 1875–1885, 2014. doi: 10.1016/j.proeng.2014.12.341.

[50] N. Vamsi Krishna Jasti and A. Sharma, "Lean manufacturing implementation using value stream mapping as a tool", *Int. J. Lean Six Sigma*, 2014. doi: 10.1108/ijlss-04-2012-0002.

[51] A. A. A. Hakeem, D. Solyali, M. Asmael, and Q. Zeeshan, "Smart Manufacturing for Industry 4.0 using Radio Frequency Identification (RFID) Technology", vol. 32, no. 1, pp. 31–38, 2020. doi: 10.17576/jkukm-2020-32(1)-05.

[52] J. S. Randhawa and I. S. Ahuja, "An investigation into manufacturing performance achievements accrued by Indian manufacturing organization through strategic 5S practices", *Int. J. Product. Perform. Manag.*, vol. 67, no. 4, pp. 754–787, 2018. doi: 10.1108/IJPPM-06-2017-0149.

[53] R. Goyat, M. K. Rai, and G. Kumar, "Recent Advances in DV-hop Localization Algorithm for Wireless Sensor Networks", in *Proceedings – 2019 3rd International Conference on Data Science and Business Analytics, ICDSBA 2019*, pp. 415–422, 2019. doi: 10.1109/ICDSBA48748.2019.00090.

[54] P. Kumar, M. Singh, and G. S. Phull, "Production lessening analysis of manufacturing unit in India: Lean Six Sigma perspective", *J. Proj. Manag.*, 2019. doi: 10.5267/j.jpm.2019.5.001.

[55] P. Tedeschi, S. Sciancalepore, A. Eliyan, and R. Di Pietro, "LiKe: Lightweight Certificateless Key Agreement for Secure IoT Communications", *IEEE Internet Things J.*, 2020. doi: 10.1109/JIOT.2019.2953549.

[56] J. Antony, M. Sony, and O. McDermott, "Conceptualizing Industry 4.0 readiness model dimensions: an exploratory sequential mixed-method study", *TQM J.*, 2021. doi: 10.1108/TQM-06-2021-0180.

[57] S. R. Pokhrel and S. Singh, "Compound TCP Performance for Industry 4.0 WiFi: A Cognitive Federated Learning Approach", *IEEE Trans. Ind. Informatics*, vol. 17, no. 3, pp. 2143–2151, 2021. doi: 10.1109/TII.2020.2985033.

[58] M. Yazdi, A. Nedjati, E. Zarei, and R. Abbassi, "A reliable risk analysis approach using an extension of best-worst method based on democratic-autocratic decision-making style", *J. Clean. Prod.*, 2020. doi: 10.1016/j.jclepro.2020.120418.

[59] S. Singaravel, J. Suykens, and P. Geyer, "Deep-learning neural-network architectures and methods: Using component-based models in building-design energy prediction", *Adv. Eng. Informatics*, vol. 38, pp. 81–90, 2018. doi: 10.1016/j.aei.2018.06.004.

[60] J. A. Garza-Reyes, J. Torres Romero, K. Govindan, A. Cherrafi, and U. Ramanathan, "A PDCA-based approach to Environmental Value Stream Mapping (E-VSM)", *J. Clean. Prod.*, 2018. doi:10.1016/j.jclepro.2018.01.121.

[61] P. Sandner, A. Lange, and P. Schulden, "The role of the CFO of an industrial company: An analysis of the impact of blockchain technology", *Futur. Internet*, vol. 12, no. 8, 2020. doi: 10.3390/FI12080128.

[62] A. Maged, S. Haridy, S. Kaytbay, and N. Bhuiyan, "Continuous improvement of injection moulding using Six Sigma: Case study", *Int. J. Ind. Syst. Eng.*, 2019. doi: 10.1504/IJISE.2019.100165.

[63] B. Wang, P. Wang, and Y. Tu, "Customer satisfaction service match and service quality-based blockchain cloud manufacturing", *Int. J. Prod. Econ.*, vol. 240, 2021. doi: 10.1016/j.ijpe.2021.108220.

[64] A. Vafadarnikjoo, H. Badri Ahmadi, J. J. H. Liou, T. Botelho, and K. Chalvatzis, "Analyzing blockchain adoption barriers in manufacturing supply chains by the neutrosophic analytic hierarchy process", *Ann. Oper. Res.*, 2021. doi: 10.1007/s10479-021-04048-6.

[65] J. Lohmer and R. Lasch, "Blockchain in operations management and manufacturing: Potential and barriers", *Comput. Ind. Eng.*, vol. 149, 2020. doi: 10.1016/j.cie.2020.106789.

[66] S. M. Idrees, M. Nowostawski, R. Jameel, and A. K. Mourya, "Security aspects of blockchain technology intended for industrial applications", *Electron.*, vol. 10, no. 8, 2021. doi: 10.3390/electronics10080951.

[67] R. Cole, M. Stevenson, and J. Aitken, "Blockchain technology: implications for operations and supply chain management", *Supply Chain Manag.*, vol. 24, no. 4, pp. 469–483, 2019. doi: 10.1108/SCM-09-2018-0309.

[68] N. Hackius and M. Petersen, "Translating High Hopes into Tangible Benefits: How Incumbents in Supply Chain and Logistics Approach Blockchain", *IEEE Access*, vol. 8, pp. 34993–35003, 2020. doi: 10.1109/ACCESS.2020.2974622.

[69] S. Saberi, M. Kouhizadeh, J. Sarkis, and L. Shen, "Blockchain technology and its relationships to sustainable supply chain management", *Int. J. Prod. Res.*, 2019. doi: 10.1080/00207543.2018.1533261.

[70] S. Kamble, A. Gunasekaran, and H. Arha, "Understanding the Blockchain technology adoption in supply chains-Indian context", *Int. J. Prod. Res.*, 2019. doi: 10.1080/00207543.2018.1518610.

[71] S. Kurpjuweit, C. G. Schmidt, M. Klöckner, and S. M. Wagner, "Blockchain in Additive Manufacturing and its Impact on Supply Chains", in *Journal of Business Logistics*, vol. 42, no. 1, pp. 46–70, 2021. doi: 10.1111/jbl.12231.

[72] W. Mougayar, "The Business Blockchain: Promise, Practice, and Application of the Next Internet Technology", *John Wiley & Sons.* p. 142, 2016. [Online]. Available: www.ddw-online.com/informatics/p320746-blockchain-technology-in-drug-discovery:-use-cases-in-r&d.html%0A www.wiley.com/en-us/The+Business+Blockchain%3A+Promise%2C+Practice%2C+and+Application+of+the+Next+Internet+Technology-p-9781119300311.

[73] V. Babich and G. Hilary, "Distributed ledgers and operations: What operations management researchers should know about blockchain technology", *Manuf. Serv. Oper. Manag.*, vol. 22, no. 2, pp. 223–240, 2020. doi: 10.1287/MSOM.2018.0752.

[74] B. Biswas and R. Gupta, "Analysis of barriers to implement blockchain in industry and service sectors", *Comput. Ind. Eng.*, vol. 136, pp. 225–241, 2019. doi: 10.1016/j.cie.2019.07.005.

[75] I. Makhdoom, M. Abolhasan, H. Abbas, and W. Ni, "Blockchain's adoption in IoT: The challenges, and a way forward", *Journal of Network and Computer Applications*, vol. 125. pp. 251–279, 2019. doi: 10.1016/j.jnca.2018.10.019.

[76] M. Lacity and S. Khan, "Exploring preliminary challenges and emerging best practices in the use of enterprise blockchains applications", in *Proceedings of the Annual Hawaii International Conference on System Sciences*, pp. 4665–4674, 2019. doi: 10.24251/hicss.2019.563.

[77] Y. Kayikci, N. Subramanian, M. Dora, and M. S. Bhatia, "Food supply chain in the era of Industry 4.0: blockchain technology implementation opportunities and impediments from the perspective of people, process, performance, and technology", *Prod. Plan. Control*, vol. 33, no. 2–3, pp. 301–321, 2022. doi: 10.1080/09537287.2020.1810757.

[78] Y. Wang, J. H. Han, and P. Beynon-Davies, "Understanding blockchain technology for future supply chains: a systematic literature review and research agenda", *Supply Chain Management*, vol. 24, no. 1. pp. 62–84, 2019. doi: 10.1108/SCM-03-2018-0148.

[79] Y. Wang, M. Singgih, J. Wang, and M. Rit, "Making sense of blockchain technology: How will it transform supply chains?", *Int. J. Prod. Econ.*, vol. 211, pp. 221–236, 2019. doi: 10.1016/j.ijpe.2019.02.002.

3 Integrating Lean Six Sigma and Industry 4.0

Ashutosh Samadhiya
and Jose Arturo Garza-Reyes

3.1 INTRODUCTION

In the decades since its inception, the lean concept has spread globally, with most organizations now using its tenets to boost productivity, quality, and consumer value while cutting costs. As a result of cutting expenses and increasing output, the lean business model ensures consistent financial results. Lean manufacturing (LM) is a set of fundamental strategies that helps reduce waste, and operations that don't add value to the product (Bhamu and Sangwan, 2014). Similarly to the "Define-Measure-Analyse-Improve-Control" (DMAIC) approach, the "Six Sigma" (SS) methodology is used to decrease process flaws (Dutta and Jaipuria, 2020). The combination of lean and six sigma, such as the Lean Six Sigma (LSS) technique, is a way to boost productivity in the workplace and, in turn, boost user satisfaction and profits (Snee, 2010). It's important to note that LM and SS are distinct but related methods (Sordan et al., 2020). The book "The Machine That Changed the World," written by Womack in 1990 to introduce the "Toyota Production System" (TPS), is largely credited for propelling LM to international prominence (Bailey et al., 2012). In addition, Motorola's SS program was created by Bill Smith in 1986 in response to the organization's recognition of the need to lessen the frequency and severity of faults while simultaneously raising the bar for product quality (Montgomery and Woodall, 2008). Throughout the last few decades, LM and SS have assisted the manufacturing sectors in achieving operational excellence, enhanced efficiency, and increased customer satisfaction. LSS is a well-known and frequently used approach that a significant number of businesses have embraced for the purpose of process improvement (Shah et al., 2008). LSS, as defined by Cherrafi et al. (2017), is an integrated strategy that represents a convergence and harmony between two strong methodologies for continuous improvement, namely LM and SS. In recent years, the LM and SS techniques have been implemented and researched together in a unified fashion (Shah et al., 2008). LSS has been ranked as one of the top models that can lead to operational excellence by several different authors (Jaeger et al., 2014).

Organizations in today's highly competitive global market have several difficulties in satisfying their consumers' wide variety of needs (Yin et al., 2017). That's why they're always looking for new ways to improve their abilities and efficiency. The goalposts on which improvement attempts are based are always shifting, yet such efforts have been around for a very long time (Salah et al., 2010). Among

DOI: 10.1201/9781003381600-3

them is LSS, which has been shown to be a reliable leader in frameworks like TPS (Salah et al., 2010). LSS offers notions, techniques, and tools that may organize production, find solutions to difficult challenges, and then enhance operations (Snee, 2010).

In recent years, with the introduction of Industry 4.0 (I4.0), a rise in the amount of connection in processes has taken place (Tissir et al., 2022). I4.0 is responsible for introducing a new management paradigm that makes use of cutting-edge technology to make processes more modular and adaptable (Rossini et al., 2019). When applied to a given industry, I4.0 has been discussed as a potential new paradigm for significantly boosting productivity via mechanization and digitization (Chiarini and Kumar, 2020). This is owed largely to the fact that cyber-physical technologies and the Internet of Things (IoT) have enabled the linking and integration of previously separate value chains and manufacturing processes (Fatorachian and Kazemi, 2018). There has been much research on I4.0 from a technical perspective (Schroeder et al., 2019). I4.0 has recently been labelled as a strategic paradigm for gaining a market advantage and bettering key performance indicators, including price, efficiency, quality, consumer satisfaction, and lead time (Bibby and Dehe, 2018). Operational Excellence (OE) approaches like LM and SS have helped businesses over the last few years achieve efficiency benefits and boost customer satisfaction with similar aims. The Integrated LSS approach, which may be thought of as the merger of the Japanese TPS (also known as "Lean" in western culture (Khan et al., 2013)) and the "American SS" (Albliwi et al., 2015), has been adopted by most of the Fortune 500. While some authors categorize LSS as a technique (Albliwi et al., 2014), others have addressed it as a top model of OE used to achieve goals analogous to I4.0 (Jaeger et al., 2014). Many different types of new or evolving technologies are used in I4.0's supply chain collaboration (Kolberg et al., 2017), and OE approaches have been used at both the organizational and supply chain levels to improve both vertical and horizontal interfaces (Tortorella et al., 2019). Although I4.0's and LSS's goal is to enhance productivity and streamline supply chains, studies examining the complementary nature of the two approaches are only getting started (Chiarini and Kumar, 2020). Therefore, in this chapter, we'll attempt to figure out whether LSS lays a solid groundwork for maximizing the advantages of I4.0 in pursuing OE. In addition, we'll explore to understand the beneficial effects of integrating LSS tools and concepts into an organization's implementation of I4.0 technologies.

3.2   LEAN SIX SIGMA (LSS)

The integration of Lean and SS approaches is what we mean when we talk about LSS. The goal of LSS is to cut down on waste and defects (variation) while simultaneously improving efficiency and output. The Lean methodology places an emphasis on effectiveness, whereas the SS approach places an equal amount of emphasis on efficiency. When used together, they are much more effective than when one procedure is carried out autonomously of the other. A thorough understanding of the functions, duties, and frameworks of the employee, as well as the prerequisites for their training, are essential to the achievement of an effective LSS deployment.

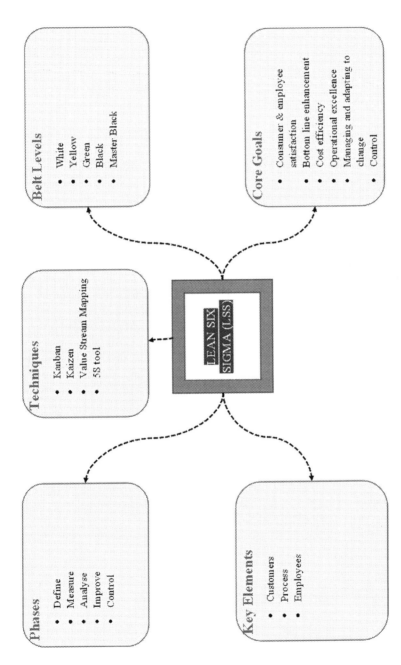

FIGURE 3.1   Lean Six Sigma summary framework (Adapted from Kenton. 2022).

3.2.1 TRACES OF LSS

The development of LSS began in 1986 in the United States at a company called Motorola. Since the end of World War II, Japan has been enjoying a financial upswing as a consequence of the application of the Kaizen business strategy. As a corollary of this, Japanese goods have traditionally been seen as being of better quality than those produced in the United States. As a result, LSS was developed as an alternative business model to combat Kaizen (DeFeo, 2019). LSS was derived from the Juran Trilogy, which is a methodology for planning, regulating, and enhancing performance inside an organization. LSS was first developed in the 1960s. The Trilogy makes use of projects with the intention of making advancements in existing degrees of performance via the use of design or enhancement strategies. Advancements do not occur by chance; rather, they call for a methodical approach to change, which is the kind of change that may be brought about by taking things "project by project" (DeFeo, 2019). As soon as a new item, service, or operation has been created or enhanced, the results need to be managed so that the benefits may be maintained. This is achieved by the use of a quality control procedure and an approach known as root cause corrective measures. The quest of bettering one's goods, operations, and services is one that never ends (DeFeo, 2019). It is possible that in order to achieve advancements, an increase of tenfold more than 3.4 parts/ million, which is the SS level, would be required. A company has to have a firm grasp on its ability to plan, manage, and enhance the quality of its goods or services if they want to continue to be innovative and fulfil the requirements of their constituents. In order to do this, they will need to make use of "Yellow Belts, Green Belts, Black Belts, and Master Black Belts" who are familiar with the LSS system (DeFeo, 2019) (see Figure 3.1).

3.3 INDUSTRY 4.0

Industry 4.0, often called the "fourth industrial revolution", is a new stage in the management and coordination of the manufacturing value chain. Industry 4.0 (including so-called "smart machines") is predicated on cyber-physical systems (IBM, n.d.). Firms are able to link and manage the operations bc through the IoT because of the advanced control modules, integrated software platforms, and Internet addresses that they possess. This allows for new modes of production, value generation, and real-time optimization by connecting goods and production methods so they may talk to one another (IBM, n.d.). I4.0 is having a profound effect on the production, improvement, and dissemination stages. Manufacturers are increasingly using cutting-edge technology in their production facilities and other parts of their businesses, such as the Internet of Things (IoT), cloud computing and analytics, artificial intelligence (AI), and machine learning.

To join I4.0, the manufacturing sector must embrace the development of smart factories. Downtime in manufacturing may be reduced significantly by using big data processing of sensor readings to monitor manufacturing resources in real time and provide predictive maintenance remedies (IBM, n.d.).

3.3.1 TRACES OF I4.0

In Britain, the "first industrial revolution" began in the late 18th century, paving the way for mass manufacturing via the use of water and steam power rather than animal and human labor. Instead of meticulously crafting each item by hand, machines were used to construct the final products. A hundred years later, the "second industrial revolution" brought about the development of assembly lines and the widespread usage of fossil fuels and modern forms of energy. Large-scale production and a certain amount of automation in the industry were made possible by these modern power supplies and the improved communications made possible by the telephone and the telegraph. Beginning in the 20th century, the "third industrial revolution" introduced new technologies such as computers, high-speed communications networks, and data analysis into production operations. In order to automate certain operations and gather and exchange data, programmable logic controllers (PLCs) were first embedded inside equipment as part of the digitalization of industries. I4.0 refers to the current phase of industrialization, which is the fourth such revolution. Data-driven decision-making is a key component of the increasingly automated, IoT-enabled, and smart-machine-reliant value chain. Ultimately, the goal is to attain efficiency with a lot size of one, which necessitates more adaptability so that producers may better satisfy client expectations via mass customization. A smart factory may improve decision-making and information openness by collecting more information directly from the factory floor and merging it with another organization's operating information (see Figure 3.2).

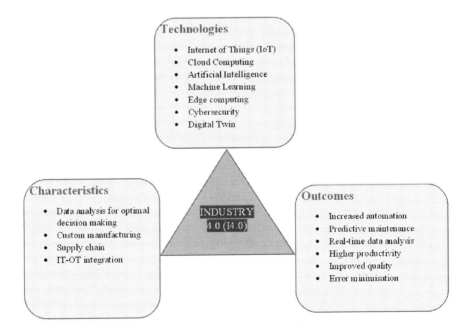

FIGURE 3.2 Industry 4.0 summary framework (Adapted from IBM, n.d.).

3.4 INTEGRATION OF I4.0 AND LSS

3.4.1 FOUNDATION OF INTEGRATING I4.0 AND LSS

The benefits of integrating I4.0 and LSS can be seen in two ways as follows:

3.4.1.1 Requirements for Implementing I4.0 Are Established by LSS

The amalgamation of I4.0 technologies may be made facilitated with the help of LSS tools and procedures by standardizing processes, cutting down on the number of errors caused by humans (Anosike et al., 2021), and lowering the amount of process variability (Chiarini and Kumar, 2020). In addition, the use of LSS methods such as VSM may assist in the picking of I4.0 technologies for a firm (Ciano et al., 2020). This is because these approaches make it easier to determine the areas in which these techniques can create maximum benefit.

3.4.1.2 While Each Is Being Deployed Independently, I4.0 and LSS Complement One Another

I4.0 technologies, for instance, have the potential to lessen the amount of work needed to keep LM operational (Rosin et al., 2019) and improve SS by making it possible to collect and process vast amounts of data in a shorter amount of time (Belhadi et al., 2020). In a similar vein, LSS has the potential to improve the performance of I4.0 as it is being implemented by identifying useful applications for the data that is gathered (through the technologies of I4.0) and making it easier to comprehend and evaluate that data (Fogarty, 2015).

Operations will be better and more automated due to I4.0, but they will still be operations (Sodhi, 2020). So long as there is a requirement for optimization, the LSS strategy will remain relevant. Several continuous improvement approaches, including LSS, have evolved due to the vast collection of novel technologies made possible by I4.0 (Arcidiacono and Pieroni, 2018). I4.0 has made significant changes, and the old LSS model will no longer apply. To face these new difficulties, LSS must be modernized to represent an intelligent and effective strategy for continuous improvement inside smart factories (Tissir et al., 2022). According to Sodhi (2020), LSS is predicated on the gathering and analysis of data to identify the main reason for a problem and the most effective ways to address it. Then, with I4.0 enablers like the IoT, businesses can quickly and efficiently collect a wide variety of information in real-time. Big data analytics (BDA) is another example of I4.0 technology; its strength comes from the combination of four characteristics: "volume, variety, velocity, and veracity". In their explanation of the benefits that BDA methods may offer to each step of the LSSs DMAIC approach, Gupta et al. (2020) highlight the great capability and attentiveness that BDA techniques can provide to the LSS. The LSS requires these methods throughout its development so that judgments may be made with more certainty and predictability as more and more data is collected and analyzed. Furthermore, Statistical Process Control is used to monitor the process and identify any unusual occurrences or fluctuations. More precise information will be gathered in live time with the help of technologies like IoT, sensors, and BD. That's why, if there's ever a problem, we'll be able to fix it right away to avoid problems.

When I4.0 technologies are amalgamated into LSS tools, production processes are improved, and the end result is more satisfying to customers (Sodhi, 2020).

3.4.2 I4.0 AND LSS: INTEGRATING FOR COMPLEMENTARY BENEFITS

Integrating the LSS method into continuous improvement with the I4.0 is gaining much attention from experts in the field. Tissir et al. (2022) focus on identifying the LSS 4.0's conceptual underpinnings from its outward manifestation and categorizing them to guide further inquiry. The study findings have been sorted into the three critical areas of "Relationship, Implementation, and Performance", as shown in the route map (Figure 3.3).

The framework proposed by Tissir et al. (2022) (see Figure 3.3) starts off with the connection between LSS and I4.0 as the starting point for change. As seen in the next section, the approaches to the relationship between paradigms vary widely. Some authors argue that the LSS method significantly simplifies the introduction of new technologies since it does away with inefficiencies and variances before resorting to robotics (Tissir et al., 2022). As a result, LSS endorses the I4.0. Some argue that I4.0 is a supporter of LSS because it facilitates the employment of cutting-edge technologies like those provided by I4.0, which in turn accelerates and derives even more advantages from the LSS method (Tissir et al., 2022). An alternative view holds that the paradigms mutually support one another, with one method compensating for the other's deficiency in achieving the goals and being highly competitive in the market. Businesses may better handle implementation difficulties and achieve their goals if they have a firm grasp on the features of the two related methods before attempting to execute them.

The framework's (see Figure 3.3) second phase involves learning how to properly incorporate LSS 4.0 into existing processes and gaining familiarity with the methodologies and tools used in implementing both strategies. To improve outcomes and enhance the likelihood of success, it is important to have clear and actionable implementation guidelines and methodologies. Thus, there is research looking for approaches, crucial success criteria, and the effect of the societal and economic setting on the rollout of the new LSS 4.0.

Key performance indicators (KPIs) must be defined to assess the degree of success in reaching the set goals. As a result, the last part of framework schematization is to develop intelligent KPIs for LSS 4.0 deployment, which will aid in maintaining and further improving the system using the Plan-Do-Check-Act (PDCA) methodology.

Based on reading and synthesis, key concepts have been identified and a conceptualization offered that follows a natural progression toward an understanding and application of the emergent combination.

3.4.3 POTENTIAL BENEFITS OF THE INTEGRATION

3.4.3.1 The Influence of I4.0 Technologies on Operational Performance May Be Optimized with the Help of a Foundation Provided by LSS

After integrating LSS and simplifying business processes to reduce variance, manufacturers may begin using I4.0 technologies. If corporations don't wait until that

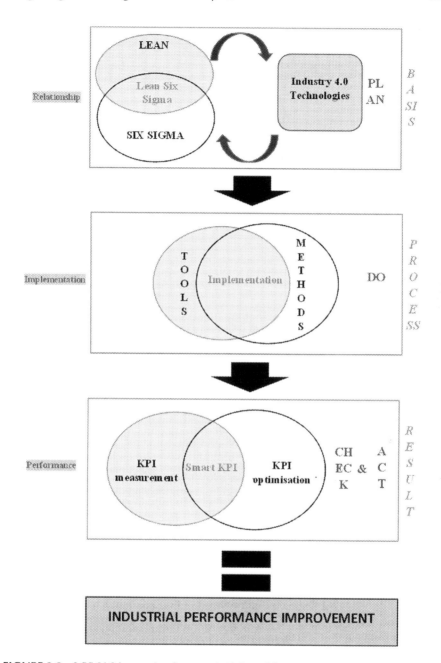

FIGURE 3.3   LSS-I4.0 integration framework (Adapted from Tissir et al., 2022).

time, they risk automating waste and increasing prices in the name of Industry 4.0. Without initially adopting LSS, businesses are free to deploy I4.0 solutions immediately. However, even though I4.0 technologies have the potential to minimize Lean waste via automation, doing so may result in less-than-ideal solutions (Chiarini and Kumar, 2020).

3.4.3.2   Improve the Efficacy of Horizontal Integration

When attempting to combine I4.0 with LSS, manufacturers should conduct process analysis using reimagined mapping tools, like an intelligent VSM and other individualized digital maps. The outcomes of this mapping process, which may aid in attaining horizontal integration, are the most operational aspect of the integration (Kamble et al., 2019). While improving all operations with cyber technology is essential, the actual end objective of horizontal integration is to accomplish "Autonomous Process Synchronization" (APS) with the help of an e-Kanban pull system, which will allow for continuous flow with a minimal quantity of stock (MacKerron et al. 2014).

3.4.3.3   Improved Process Mapping

From the perspective of forecasting and real-time supervision, digital technology may supplement conventional process mapping methods like the flowchart, VSM, and swimlane chart.

3.4.3.4   Performance Measuring

The ability to supervise key performance indicators (KPIs) in actual time is made possible through the use of sensors, actuators, IoT, radio frequency identification (RFID), and machine-to-machine communication (M2M).

3.4.3.5   Instrumentation for Assessing Machine Health

The performance of manufacturing businesses' equipment and its maintenance is crucial to their reliability (Samadhiya et al., 2022). E-maintenance, enabled by RFID, BDA, and IIoT, will replace conventional maintenance practices within the framework of Total Productive Maintenance (TPM), allowing for continuous monitoring of parts and equipment (Li et al., 2015).

3.4.3.6   Shorten the Time Required for Setting Up

There will be less time spent on setup since the raw materials can converse straight to the processing machinery using machine learning, RFID, and self-optimization (Sanders et al., 2016). To that end, additive manufacturing might print replacement components and machine pieces that plug into existing systems (Sanders et al., 2017).

3.4.3.7   Production Monitoring in Real-Time

The visual method known as "Andon" is often used to aid in applying the Jidoka. Its purpose is to update employees on the process's current state and increase their sense of urgency. When used to CPS (Cyber-Physical Systems), some sensors will be equipped to detect and rectify defects automatically, sending alerts to operators' wearable technology in real time (Kolberg and Zühlke, 2015). The Andon scheme offers an immediate insight into the operation performance, such as per-unit tracking and web analytics for the factory floor, to help in production supervision (Powell et al., 2018).

REFERENCES

Albliwi, S. A., Antony, J., & Lim, S. A. (2015). A systematic review of Lean Six Sigma for the manufacturing industry. Business Process Management Journal, 21(3), 665–691.

Albliwi, S., Antony, J., Lim, S.A. , & Van der Wiele, T. (2014). Critical failure factors of Lean Six Sigma: A systematic literature review. International Journal of Quality & Reliability Management, 31(9), 1012–1030.

Anosike, A., Alafropatis, K., Garza-Reyes, J. A., Kumar, A., Luthra, S., & Rocha-Lona, L. (2021). Lean manufacturing and Internet of things – A synergetic or antagonist relationship? Computers in Industry, 129, 103464.

Arcidiacono, G., & Pieroni, A. (2018). The revolution Lean Six Sigma 4.0. International Journal on Advanced Science Engineering Information Technology, 8(1), 141–149.

Bailey, A. Y., Motwani, J., & Smedley, E. M. (2012). When lean and Six Sigma converge: A case study of a successful implementation of Lean Six Sigma at an aerospace company. International Journal of Technology Management, 57(1/2/3), 18.

Belhadi, A., Kamble, S. S., Zkik, K., Cherrafi, A., & Touriki, F. E. (2020). The integrated effect of big data analytics, Lean Six Sigma and green manufacturing on the environmental performance of manufacturing companies: The case of North Africa. Journal of Cleaner Production, 252, 119903.

Bhamu, J., & Sangwan, K. S. (2014). Reduction of post-kiln rejections for improving sustainability in ceramic industry: A case study. Procedia CIRP, 26, 618–623.

Bibby, L., & Dehe, B. (2018). Defining and assessing industry 4.0 maturity levels – case of the defence sector. Production Planning & Control, 29(12), 1030–1043.

Cherrafi, A., Elfezazi, S., Chiarini, A., Mokhlis, A., & Benhida, K. (2017). The integration of lean manufacturing, Six Sigma and sustainability: A literature review and future research directions for developing a specific model. Journal of Cleaner Production, 139, 828–846.

Chiarini, A., & Kumar, M. (2020). Lean Six Sigma and industry 4.0 integration for operational excellence: Evidence from Italian manufacturing companies. Production Planning & Control, 32(13), 1084–1101.

Ciano, M. P., Dallasega, P., Orzes, G., & Rossi, T. (2020). One-to-one relationships between industry 4.0 technologies and lean production techniques: A multiple case study. International Journal of Production Research, 59(5), 1386–1410.

DeFeo, J. A. (2019). Lean Six Sigma, lean & Six Sigma: A definitive guide. Juran. Accessed on 20th December 2022 from www.juran.com/blog/guide-to-lean-and-lean-six-sigma/

Dutta, S., & Jaipuria, S. (2020). Reducing packaging material defects in beverage production line using Six Sigma methodology. International Journal of Six Sigma and Competitive Advantage, 12(1), 59.

Fatorachian, H., & Kazemi, H. (2018). A critical investigation of industry 4.0 in manufacturing: Theoretical operationalisation framework. Production Planning & Control, 29(8), 633–644.

Fogarty, D. J. (2015). Lean six sigma and big data: continuing to innovate and optimize business processes. Journal of Management and Innovation, 1(2), 2–20.

Gupta, S., Modgil, S., & Gunasekaran, A. (2020). Big data in lean six sigma: A review and further research directions. International Journal of Production Research, 58(3), 947–969.

IBM (n.d.). What is industry 4.0 and how does it work? IBM – United States. Accessed on 20th December 2022 from www.ibm.com/in-en/topics/industry-4-0

Jaeger, A., Matyas, K., & Sihn, W. (2014). Development of an assessment framework for operations excellence (OsE), based on the paradigm change in operational excellence (OE). Procedia CIRP, 17, 487–492.

Kamble, S., Gunasekaran, A., & Dhone, N. C. (2019). Industry 4.0 and lean manufacturing practices for sustainable organisational performance in Indian manufacturing companies. International Journal of Production Research, 58(5), 1319–1337.

Kenton, W. (2022). Lean Six Sigma Explained: Definition, Principles, Tools & Belts. Investopedia. Accessed on 20th December 2022 from www.investopedia.com/terms/l/lean-six-sigma.as

Khan, M. S., Al-Ashaab, A., Shehab, E., Haque, B., Ewers, P., Sorli, M., & Sopelana, A. (2013). Towards Lean Product and Process Development. International Journal of Computer Integrated Manufacturing 26(12), 1105–1116.

Kolberg, D., Knobloch, J., & Zühlke, D. (2017). Towards a lean automation interface for workstations. International Journal of Production Research, 55(10), 2845–2856.

Kolberg, D., & Zühlke, D. (2015). Lean automation enabled by industry 4.0 technologies. IFAC-PapersOnLine, 48(3), 1870–1875.

Li, J., Tao, F., Cheng, Y., & Zhao, L. (2015). Big data in product lifecycle management. The International Journal of Advanced Manufacturing Technology, 81(1–4), 667–684.

MacKerron, G., Kumar, M., Kumar, V., & Esain, A. (2014). Supplier replenishment policy using E-kanban: A framework for successful implementation. Production Planning & Control, 25(2), 161–175.

Montgomery, D. C., & Woodall, W. H. (2008). An overview of six sigma. International Statistical Review, 76(3), 329–346.

Powell, D., Romero, D., Gaiardelli, P., Cimini, C., & Cavalieri, S. (2018). Towards digital lean cyber-physical production systems: industry 4.0 technologies as enablers of leaner production. IFIF International Conference on Advances in Production Management Systems, 353–362.

Rosin, F., Forget, P., Lamouri, S., & Pellerin, R. (2019). Impacts of industry 4.0 technologies on lean principles. International Journal of Production Research, 58(6), 1644–1661.

Rossini, M., Costa, F., Tortorella, G. L., & Portioli-Staudacher, A. (2019). The interrelation between industry 4.0 and lean production: An empirical study on European manufacturers. The International Journal of Advanced Manufacturing Technology, 102(9–12), 3963–3976.

Salah, S., Rahim, A., & Carretero, J. A. (2010). The integration of Six Sigma and lean management. International Journal of Lean Six Sigma, 1(3), 249–274.

Samadhiya, A., Agrawal, R., & Garza-Reyes, J. A. (2022). Integrating industry 4.0 and total productive maintenance for global sustainability. The TQM Journal. https://doi.org/10.1108/tqm-05-2022-0164\

Sanders, A., Elangeswaran, C., & Wulfsberg, J. (2016). Industry 4.0 implies lean manufacturing: Research activities in industry 4.0 function as enablers for lean manufacturing. Journal of Industrial Engineering and Management, 9(3), 811.

Sanders, A., Subramanian, K. R., Redlich, T., & Wulfsberg, J. P. (2017). Industry 4.0 and lean management–synergy or contradiction? IFIP International Conference on Advances in Production Management Systems, 341–349.

Schroeder, A., Ziaee Bigdeli, A., Galera Zarco, C., & Baines, T. (2019). Capturing the benefits of industry 4.0: A business network perspective. Production Planning & Control, 30(16), 1305–1321.

Shah, R., Chandrasekaran, A., & Linderman, K. (2008). In pursuit of implementation patterns: The context of lean and Six Sigma. International Journal of Production Research, 46(23), 6679–6699.

Snee, R. D. (2010). Lean Six Sigma – getting better all the time. International Journal of Lean Six Sigma, 1(1), 9–29.

Sodhi, H. (2020). When Industry 4.0 meets Lean Six Sigma: A review. Industrial Engineering Journal, 13(1), 0–12.

Sordan, J. E., Oprime, P. C., Pimenta, M. L., Chiabert, P., & Lombardi, F. (2020). Lean Six Sigma in manufacturing process: A bibliometric study and research agenda. The TQM Journal, 32(3), 381–399.

Tissir, S., Cherrafi, A., Chiarini, A., Elfezazi, S., & Bag, S. (2022). Lean Six Sigma and industry 4.0 combination: Scoping review and perspectives. Total Quality Management & Business Excellence, 1–30.

Tortorella, G. L., Giglio, R., & Van Dun, D. H. (2019). Industry 4.0 adoption as a moderator of the impact of lean production practices on operational performance improvement. International Journal of Operations & Production Management, 39(6/7/8), 860–886.

Yin, Y., Stecke, K. E., & Li, D. (2017). The evolution of production systems from industry 2.0 through industry 4.0. International Journal of Production Research, 56(1–2), 848–861.

4 Mapping Fourth Industrial Revolution Enabling Technologies

Adriano Gomes de Freitas,
Catarina de Andrade Lucizano,
Alexandre Acácio de Andrade,
and Júlio Francisco Blumetti Facó

4.1 INTRODUCTION

The Fourth Industrial Revolution is a widely spread concept among different industries: manufacturing, services, construction, education, etc., and its development is related to the changes in consumer needs all over the world. Following its predecessors (First, Second, and Third Industrial Revolutions), the Fourth Industrial Revolution, also known as Industry 4.0, arises as a response to today's consumption patterns, where industries must respond quickly to requests for highly customizable, environmentally friendly, and cost-effective products (KELLER et al., 2014; KHALID et al., 2017; RÜSSMANN, 2015). For the manufacturing industries, the challenge starts as early as during the product's idealization: how is it possible to constantly add compelling features to already well-established products? Furthermore, how to create means of production flexible enough to be able to produce small scales of unlimited versions of personalized goods? There are also concerns about how cost-competitive nations can take advantage of this disruption to unsettle the ones that are historically production leaders (DAUDT; WILLCOX, 2016; CUCINOTTA et al., 2009; PICCININI et al., 2015).

To tackle that demand and be able to expand, the common trend is to deploy digital technologies that will provide industries with more flexibility and agility. In fact, the three pillars upon which the Fourth Industrial Revolution rests are flexibility, competitiveness, and sustainability (KELLER et al., 2014). While the objective is fairly clear and there's a certain level of common sense when it comes to what are the relevant technologies to achieve Industry 4.0 goals, the path to deploying them can be ambiguous and to effectively initiate the digitization process is not obvious. In fact, according to the German Federal Ministry of Education and Research, there's a need for fuller and more targeted information when it comes to how to implement Industry 4.0 principles (BACKES-GELLNER et al., 2013).

DOI: 10.1201/9781003381600-4

4.1.1 OBJECTIVES

This study's objective is to propose a systemic roadmap for the implementation of Industry 4.0 enabling technologies, based on the comparison of the necessary and attainable capabilities by each technology at different levels of maturity.

4.2 LITERATURE REVIEW

The emergence of the Fourth Industrial Revolution is marked by a disruption from the dynamics of its three predecessors: instead of being first observed as it develops in an almost organic way, the Fourth Revolution was foreseen and planned by governments, universities, and private initiatives (HERMANN et al., 2015).

In the face of growing international competition, several countries have launched structured programs for the development and application of new technologies in industrial environments. The first nation to publicly do it was Switzerland, in 2008 (KAGERMANN; WAHLSTER; HELBIG, 2013). However, the expression "Industry 4.0" was only publicly introduced in 2011 in Germany at the annual Hannover Fair. Due to its long history of being one of the most innovative countries, but also facing its greatest economic crisis since the WW2, Germany's Federal Ministry of Education and Research released in 2013 a document elaborated by the Industry 4.0 Working Group with recommendations for implementing the new strategic initiative and it became the cornerstone for the following research and development of this new industrial paradigm (BACKES-GELLNER et al., 2013 LIDONG; GUANGHUI, 2016).

The Fourth Industrial Revolution principles for the manufacturing industry, or Industry 4.0 principles, are:

- Service orientation: the functionalities of the systems must be able to be offered as services.
- Interoperability: physical, virtual, and human systems must be able to communicate transparently.
- Virtualization: physical systems must be equipped with resources that allow their transcription to the virtual world.
- Decentralization: individual systems can make autonomous decisions, waiving the need for upper management commands.
- Modularization: design of systems that allow rapid adaptation in the face of constantly changing requirements.
- Synchronicity: ability to collect, process and deliver information and actions in real time.

Among the benefits promoted by Industry 4.0, the most often mentioned by manufacturing managers is the intensification of competitiveness. Changes in companies' strategies combined with the adoption of new technologies foster the need for new partnerships, in addition to improving local performance to meet customer demands. Thus, business opportunities flourish and increased competitiveness spreads throughout the production chain (KIEL et al., 2017).

However, there is a lack of deep understanding of how to start the so-called "digitization process" and the precise knowledge of its steps is a determining factor as to whether an industry will fail or succeed in implementing it.

4.3 METHODS

In 2020, the German National Academy of Science and Engineering published the *Industrie 4.0 Maturity Index*, where a 6-stage maturity model is proposed considering capabilities displayed by companies individually (SCHUH et al., 2020), and this index will be used in this work as a guide to collate the identified Industry 4.0 enabling technologies. However, the index itself does not explicitly link technologies to each level of maturity. Therefore, to achieve the main objective of this work, four interim steps will be pursued:

1. Identification of the technologies that are considered enabling for the Fourth Industrial Revolution.
   - On this step, bibliographic research will be conducted to determine what are the technologies considered as enabling Industry 4.0.
2. Mapping the technical dependencies among the elected technologies.
   - Once the enabling technologies have been selected, a detailed exploration of their definitions, characteristics, and usage will be conducted. A deeper understanding of how these technologies operate allows the identification of technical dependencies among them and makes it possible to establish an implementation sequence based on technical premises.
3. Maturity index analysis and identification of what are the competencies that the elected technologies aim to enable.
   - The Industry 4.0 Maturity Index proposed by the ACATECH will be used as the source that determines what are the non-technical competencies that a company needs to display to be engaged to the I4.0 principle in different levels of maturity.
   - Taking hold of the previous step outcome, the non-technical competencies fostered by the adoption of each enabling technology will be mapped and the technologies distributed at each maturity level that they may be more adherent to.
4. Proposal of an implementation roadmap that is technically coherent and complies with the desired maturity evolution.
   - On this step, the studies previously conducted will be overlapped to verify at what level the technical sequence and the maturity level distribution will converge.

4.3.1 IDENTIFICATION OF THE INDUSTRY 4.0 ENABLING TECHNOLOGIES

There are no official international norms or standards established as to which technologies are considered enablers for Industry 4.0. However, there is a certain level of agreement between academic works (BORTOLINI; GALIZIA; MORA, 2018; CHEAH; LEONG, 2019; NAKAYAMA, 2017; SATURNO et al., 2018) and other studies regarding this topic, such as those published by Capgemini, McKinsey and Boston Consulting Group (BECHTOLD et al., 2014; MCKINSEY, 2015; RÜSSMANN, 2015).

Table 4.1 displays the technologies cited by authors whose work specifically addresses the indication of Industry 4.0 enabling technologies. It also shows the frequency with which each technology is addressed by different authors. In total, 16 different technologies were proposed, four of which were only addressed once: IoS, cognitive computing, RFID, and Smart ERPs; and one cited twice: Mobile Devices. These, as they represent less than 30% of the frequency of approach by the studied authors, were excluded from the present analysis. Thus, the technologies here considered as enabling for the Fourth Industrial Revolution are:

- Industrial Internet of Things (IIoT).
- Cyber Physical Systems (CPS).
- Digital simulations.
- Cyber security.
- Additive manufacturing.
- Collaborative robots.
- Big Data Analytics.
- Augmented reality.
- Horizontal and Vertical Systems Integration (HVSI).
- Cloud computing.
- Smart Sensors.

4.3.2 Industry 4.0 Maturity Index Analysis

The maturity assessment on how a company is complying with the Fourth Industrial Revolution premises results in its classification at one of the six different levels proposed by the German National Academy of Science and Engineering. This indicator is based on the available infrastructures and technologies in addition to the analysis of their corporate strategies and processes. It can provide an outlook on the status and on what should be pursued by the company to achieve higher levels of maturity if that's an objective.

As shown in Figure 4.1, each maturity level represents a capability to be conquered by the company as it moves towards full adherence to Industry 4.0: Visibility, Transparency, Predictive Capability and Adaptability. However, the document does not link specific technologies to each level, except for the Transparency level, which the document associates to Big Data.

- Visibility: at this level, the adoption of devices capable of capturing and providing in real time the status of each of the company's processes stands out, guaranteeing the visibility of these processes. The security of the data is also an important concern.
- Transparency: the outcome of the massive amount of data previously collected after being treated, grouped and contextualized.

TABLE 4.1
Industry 4.0 Enabling Technologies

	Authors						
	Bortolini et al., 2018	Cheah et al., 2019	Nakayama, 2017	Saturno et al., 2018	Capgemini, 2014	McKinsey, 2015	BCG, 2015
IoT	X	X	X	X		X	X
IoS			X				
Smart ERP			X				
Big Data	X	X		X	X	X	X
Additive Manufacturing	X	X		X	X	X	X
Cloud Computing	X	X		X	X	X	X
Cyber Security	X	X		X			X
Collaborative Robots		X		X	X		X
Augmented Reality	X	X				X	X
Digital Simulations	X	X				X	X
Cognitive Computing				X			
Mobile Devices				X	X		
RFID				X			
Cyber-Physical Systems	X				X	X	X
Smart Sensors	X			X	X	X	
Horizontal and Vertical Systems Integration	X	X					X

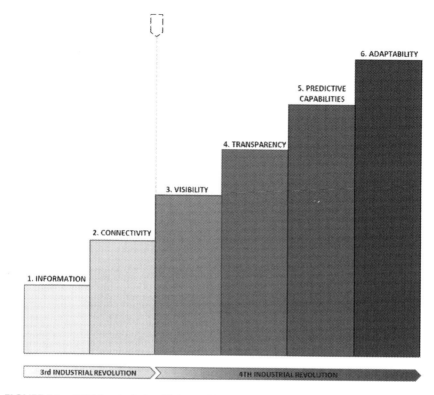

FIGURE 4.1 I4.0 Maturity Index (Elaborated by the authors and adapted from Schuh et al., 2018).

- Predictive Capability: here, the company should be able to apply tools to predict deviations in the process before they occur.
- Adaptability: the process is robust enough to operate autonomously. It is possible that adaptations to sudden events, be they operational failures, production scope changes, etc., occur as quickly as possible, preferably in real time.

4.4 RESULTS AND DISCUSSIONS

4.4.1 CORRELATIONS BETWEEN ENABLING TECHNOLOGIES OF INDUSTRY 4.0 AND MATURITY LEVELS

After the Industry 4.0 enabling technologies were reviewed, it was noted that they do have technical correlations to each other, whether direct or given through other technologies. These links were mapped and the ones that are characterized as direct technical dependents – that is, when one technology requires another to fully function – were established. Ultimately, it was possible to group the technologies to have an overview of these relationships as shown in Figure 4.2.

Once the redundancies are removed, it's possible to re-align the technologies, as shown on Figure 4.3 and obtain a more accurate perception of what the implementation order would be, based only on technical aspects.

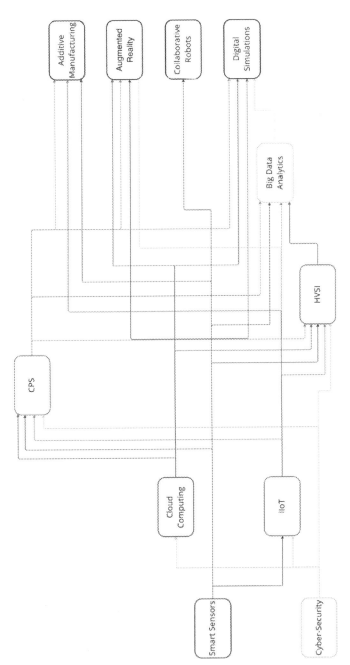

FIGURE 4.2 Technical correlations between the enabling technologies.

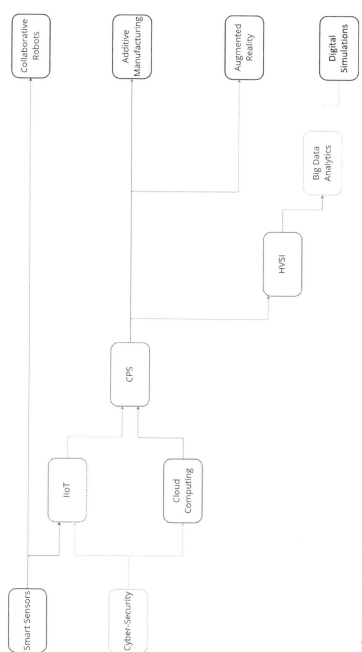

FIGURE 4.3 Simplified technical correlations between the enabling technologies.

Although it already displays some level of ordination, it is not possible to take the technology sequence displayed in Figure 4.3 as conclusive because it does not take into consideration the non-technical capabilities discussed previously. The maturity index is composed of both technical and non-technical aspects; therefore, it is necessary to correlate the capabilities leveraged by the technologies individually to the capabilities required by each maturity level.

Table 4.2 displays the correlation between the competencies for each maturity level and the technologies that enable them, based on the previous exploration of its characteristics. Only the maturity levels concerning the Fourth Industrial Revolution were listed.

Once each enabling technology is associated with the maturity level, it's possible to establish a systemic roadmap that takes into consideration both technical and non-technical aspects of what is expected from the Fourth Industrial Automation concept.

Figure 4.4 displays the result of this study. The model is proposed in levels for two fundamental reasons. The first concerns helping the planning process, so that there is a clear objective: to achieve the competence proposed by the level; as well as the technical path to be adopted to achieve this objective: the sequence of technologies. In this format, the establishment of resources and metrics is favored. The second reason is to highlight the need to assess the impact that the implementation of these technologies has generated, if not individually, in the general context of the level. This assessment is an important decision-making tool, as it will indicate whether the performance of the implemented technologies is positive and whether the competencies from a strategic point of view have been achieved, thus indicating whether the company is prepared to advance to higher levels.

TABLE 4.2
Correlation Between the Competencies for Each I4.0 Maturity Level

Maturity Level	Competencies	Enabling Technologies
1. Visibility	Capturing and providing real time data	Smart Sensors IIoT Cyber Security
2. Transparency	Full vision of the data, after it's been contextualized	Big Data Vertical and Horizontal Systems Integration Cloud Computing Cyber physical Systems.
3. Predictive Capability	Predict deviations in the process before they occur.	Digital Simulations Augmented Reality
4. Adaptability	Adaptations should occur as quickly as possible	Additive Manufacture Collaborative Robots

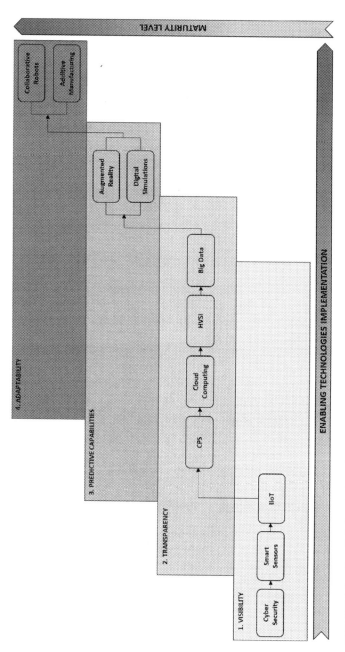

FIGURE 4.4 Detailed description of enabling technologies.

4.5   CONCLUSION

This work aimed to investigate the relationship between the technologies associated with Industry 4.0 and the strategic competencies that an industry in compliance with it must display. A set of eleven technologies was identified, among the ones appointed by authors both in the academic and corporate spheres and were considered as the enablers of this new model. It was also observed that most of these technologies have technical interdependencies, which naturally imposes a logical order of implementation. As for the strategic competencies, the ACATECH model of industry maturity assessment was used as a basis for relating the functionalities provided by each technology to the competencies expected in each of the maturity levels. The study accomplishes its objective of establishing a systemic roadmap for the implementation of Industry 4.0 enabling technologies, however, there are considerations that should be made about its context. The model considers that the industry at issue should be at the highest level of maturity regarding the Third Industrial Revolution, no technical requirements were established prior to the first level of maturity for the Fourth Industrial Revolution. Also, the model's adherence should be tested on a concrete case to confirm whether the theoretical inferences made are applicable.

ACKNOWLEDGMENTS

The authors would like to thank all the people from Federal University of ABC, University of São Paulo, University of Bradford and Monash University for their support that made this project possible.

FUNDING

The authors would like to thank the Brazilian National Council for Scientific and Technological Development (CNPq) together with the Federal University of ABC, under Grants 140368/2018-3 (Doctorate), 163815/2018-6 (Energy), and 200716/2020-4 (exchange at Monash University). Co-funded by the Erasmus+ program of the European Union, at The University of Bradford/UK under the UB Number: 21042490.

DECLARATION OF INTEREST STATEMENT

According to the publisher policy and my researcher's ethical obligation, I declare that I received funding from CNPq, Erasmus+ program. I have disclosed those interests fully to the publisher, and I have in place an approved plan for managing any potential conflicts arising from that financial support.

JEL CODE

L80   Industry Studies: Services: General
L89   Industry Studies: Services: Other
O00   Economic Development, Innovation, Technological Change, and Growth

O30 Innovation; Research and Development; Technological Change;
O31 Innovation and Invention: Processes and Incentives
O32 Management of Technological Innovation and R&D
O33 Technological Change: Choices and Consequences; Diffusion Processes
O35 Social Innovation
O36 Open Innovation

REFERENCES

BACKES-GELLNER, Uschi et al. *Research, Innovation and Technological Performance in Germany* (Report 2013). SSRN Electronic Journal, p. 160, 2013.

BECHTOLD, Jochen et al. *Industry 4.0 – The Capgemini Consulting View.* Capgemnini Consulting, p. 31, 2014.

BORTOLINI, Marco; GALIZIA, Francesco Gabriele; MORA, Cristina. Reconfigurable manufacturing systems: Literature review and research trend. *Journal of Manufacturing Systems,* v. 49, n. October, p. 93–106, 2018. Available in: https://doi.org/10.1016/j.jmsy.2018.09.005.

CHEAH, Sin-Moh; LEONG, Helene. *Relevance of Cdio To Industry 4.0-Proposal for 2 New Standards.* 2019. Available in: www.aethon.com.

CUCINOTTA, Tommaso et al. A real-time service-oriented architecture for industrial automation. *IEEE Transactions on Industrial Informatics,* v. 5, n. 3, p. 1–11, 2009.

DAUDT, Gabriel Marino; WILLCOX, Luiz Daniel. *Reflexões críticas a partir das experiências dos Estados Unidos e da Alemanha em manufatura avançada.* BNDES Setorial, n. 44, p. 5–45, 2016. Available in: <https://web.bndes.gov.br/bib/jspui/handle/1408/9936>.

HERMANN, M., PENTEK, T. and OTTO, B. Design principles for Industrie 4.0 scenarios: a literature review. Technische Universität Dortmund, Dortmund, 45, 2015.

KAGERMANN, Henning, WAHLSTER, Wolfgang, HELBIG, Johannes. *Germany – INDUSTRIE 4.0. Final report of the Industrie 4.0* WG, n. April, p. 82, 2013.

KELLER, Michael et al. How Virtualization, Decentralization and Network Building Change the Manufacturing Landscape: An Industry 4.0 Perspective. *International Journal of Mechanical, Aerospace, Industrial, Mechatronic and Manufacturing Engineering,* v. 8, n. 1, p. 37–44, 2014.

KHALID, A et al. Implementing Safety and Security Concepts for Human-Robot Collaboration in the context of Industry 4.0 Towards Implementing Safety and Security Concepts for Human-Robot- Collaboration in the context of Industry 4.0. *International MATADOR Conference on Advanced Manufacturing,* n. July, p. 1–7, 2017.

KIEL, Daniel, et al. Sustainable Industrial Value Creation: Benefits and Challenges of Industry 4.0. *International Journal of innovation management,* v 21, n 8, 2017.

LIDONG, Wang, GUANGHUI, Wang. Big Data in Cyber-Physical Systems. Digital Manufacturing and Industry 4.0. International Journal of Engineering and Manufacturing, v. 6, n. 4, p. 1–8, 2016.

MCKINSEY. *Industry 4.0 How to navigate digitization of the manufacturing sector.* McKinsey Digital, n. 1, p. 62, 2015.

NAKAYAMA, R U Y Somei. *Oportunidades de atuação na cadeia de fornecimento de sistemas de automação para indústria 4.0 no Brasil.* 2017. 240 f. São Paulo Federal University, 2017.

PICCININI, Everlin et al. Transforming industrial business: The impact of digital transformation on automotive organizations. *2015 International Conference on Information Systems: Exploring the Information Frontier, ICIS 2015,* n. September, 2015.

RÜSSMANN, Michael et al. Industry 4.0: Future of Productivity and Growth in Manufacturing. 2015. Available in :< www.bcg.com/ptbr/publications/2015/engineered_products_pro-ject_business_industry_4_future_productivity_growth_manufacturing_industries.aspx.

SATURNO, M. et al. Proposal of an Automation Solutions Architecture for Industry 4.0. *DEStech Transactions on Engineering and Technology Research*, n. icpr, **2018**.

SCHUH, Günther et al. Industrie 4.0 Maturity Index. Managing the Digital Transformation of Companies. *Web*, v. 1, n. 5765, p. 46, 2020. Available in: <www.acatech.de/publikatio nen.>.

5 Lean, Six Sigma, and Industry 4.0 Technologies Connection and Inclusion

Geeta Sachdeva

5.1 INTRODUCTION

Industry 4.0 is a new manufacturing system that brings about notable gains because of evolving operating framework conditions. Industry 4.0 (I4.0) has been discussed as a paradigm for drastically boosting production with the help of computerization & digitalization. This is mostly because of the linking & integrating of value chains and industrial systems via cyber-physical-space and Internet of Things (IoT) technologies (Ghobakhloo, 2018) I4.0 has mainly been explored from a technology viewpoint since its inception. It has been categorized throughout the past five years as a strategy paradigm for gaining competitive advantage and improving KPIs like cost, productivity, quality, customer happiness, and lead time (Kolberg et al., 2017; Agrifoglio et al., 2017; Lu, 2017).

Operational Excellence techniques like Lean and Six Sigma (LSS) have helped firms over the past three decades increase productivity & boost client gratification. The Integrated Lean Six Sigma approach that could be thought of as the merger of the Japanese Toyota Production System (TPS), has been adopted by the majority of Fortune 500 businesses.

Eliminating waste or activities with no added value from manufacturing processes is known as the "Toyota way of production" or lean manufacturing. The key principles of Lean have been adopted and implemented by the majority of organizations in order to increase efficiency and quality while lowering costs and creating value for customers. The Lean concept was first introduced in Japan decades ago. Lean business models improve productivity, and reduce costs, to stabilize financial performance. Lean manufacturing is a mix of fundamental techniques that reduces waste and operations that don't provide value (Bhamu and Sangwan, 2014). Another tool for quality improvement that uses "Define-Measure-Analyze-Improve-Control" (DMAIC) to decrease process flaws is called Six Sigma (Dutta and Jaipuria, 2020;

DOI: 10.1201/9781003381600-5

Karnjanasomwong and Thawesaengskulthai, 2019). In the modern day, manufacturing firms are moving toward innovative production concepts to boost productivity and consumer happiness. I4.0 is the process of implementing cutting-edge info systems & technologies to improve manufacturing. These include the Industrial Internet of Things (IIoT), robotics, & artificial intelligence (AI). Continuous improvement approaches like Lean Six Sigma have found increased use as Industry 4.0 has changed how manufacturing systems operate, resulting in what is described as the Fourth Industrial Revolution. I4.0 has emerged as a practical strategy that drives the manufacturing process toward meeting the requirements of each client (Vaidya et al., 2018). It combines a variety of smart technologies, including cloud computing, block chain, additive manufacturing & data analytics (Bakator et al., 2019), as well as Cyber-Physical Systems (CPS), the Internet of Things (IoT), Enterprise Architecture (EA), enterprise integration, and information and communication technology. Due to the availability of sufficient funding, large firms are integrating these technologies into their organizations, but small size industries are lagging behind and are reluctant to adapt.

Innovative I4.0 technology supports Lean Six Sigma's (or Lean and Six Sigma's separately) data-driven methodologies. Process optimization, particularly in data gathering and analysis, can help firms better utilize cutting-edge technologies. In order to enhance operational effectiveness, Shahin et al. (2020) studied the association between popular Lean tools and I4.0 technologies & conducted a case study on cloud-based Kanban. Due to the constant changes in consumer preferences, globalization, and market demands, small and medium-sized businesses (SMEs) face greater difficulties; thus, there is a critical need to connect Lean thinking with I4.0 technology.

In order to accomplish the fundamental goal of boosting efficiency and flexibility in the production system, Lean manufacturing and I4.0 work in tandem.

Industries will be given a way, by the integration of these, to provide quality products with additional worth to consumers.

The present chapter's major goal is to look into the relationships between Lean, Six Sigma, and I4.0, regardless of the kind of organization. There are only a small number of studies that discuss the amalgamation of Six Sigma with I4.0 technologies (Biswas, 2019; Sodhi, 2020; etc.). In order to give extraordinary value to customers, it is crucial to integrate the Lean and Six Sigma approach as the state of production today shifts toward smart manufacturing. The current study is based on qualitative research, and it explores the existing perspective of Lean & Six Sigma in the perspective of I4.0 as well as the interplay between these production technologies.

5.2 LITERATURE REVIEW

5.2.1 BACKGROUND ON LEAN SIX SIGMA TOOLS AND PRINCIPLES

This segment's goal is to organize the key Lean Six Sigma concepts and tools into categories so that they can be compared to Industry 4.0 technology.

A first Value Stream Mapping (VSM) activity, which is a specialized process map for identifying all waste including cost of poor quality (COPQ), is where Lean & Six Sigma projects would originate, as per some writers (Salah et al., 2010; Snee, 2010).

A future state map is created from the study of the existing state map, and an implementation plan including Kaizen waste reduction projects is then created.

Kaizen events are improvement ingenuities carried out by operational kaizen teams using the TPS's integrated tools and guiding principles (Cheng, 2018). The seven TPS wastes – "overproduction, inventory, transportation, motion, defect, waiting, and over processing" are reduced because of the application of Lean technologies, according to Ohno (1988).

5.2.2 LEAN TOOLS AND PRINCIPLES AND THEIR PURPOSE

VSM maps the processes' current condition and suggests a forthcoming state for preventing wastage. **Transactional processes** are mapped out using Lean principles to cut down on lead times and waste (Monden, 2011). **Lean Metrics** are a set of visual KPIs used in management that are related to lead-time and waste at all levels & across all departments. It creates a visual system for managing shop-floor performance on a daily basis (Khadem et al., 2008). **JIT (Just-In-Time)** only produces goods when there is a market for them, guaranteeing a constant flow from raw materials to completed goods and coordinating every operation (Monden, 2011). **Production levelling** involves balancing operations in accordance with the rhythm of orders, levelling orders while avoiding huge lots (Matzka et al., 2012). Cleaning and organizing the workspace, including the supplies, equipment, gauges, etc., uses the 5S method. Visual management, material flow management, and all other tools are built on the principles of 5S (Al-Aomar, 2011). Designing a certain layout and assembling workstations in a sequential manner to facilitate a levelled flow of materials with the least amount of transport or delay is known as "cellular manufacturing" (Salum, 2000). Usually U-shaped, the cell attempts to create the so-called one-piece flow. Single Minute Exchange of Die (SMED) cuts down on machine setup time and eliminates large lots in the process (Shingo, 1996).

Implementing automatic methods to identify issues with machinery while giving workers the option of fixing the issue or shutting down the machine is known as "Jidoka Automation" (Baudin, 2007). Making a visual cue to only activate the manufacture of components from upstream workstations when necessary & in the appropriate amounts is known as Kanban (Powell, 2018).

Six Sigma is a statistical technique grounded on the "Define, Measure, Analyze, Improve, and Control" (DMAIC) methodology for measuring & analyzing processes to eliminate variability. It is a data-driven procedure that is used to reduce variances in manufacturing processes and prevent the development of errors. Big data techniques, that gather accurate and useful data, help to properly build Six Sigma objectives (Mendonca et al., 2018). Six Sigma improves visibility in fault finding and issue solving when assisted by digital technologies.

5.2.3 BACKGROUND ON INDUSTRY 4.0 TECHNOLOGIES

The Internet of Things (IoT), cloud services, big data, and analytics are just a few of the foundational technologies used in the organized and complex model known as

Industry 4.0 to supply real-time data to production and service systems for analytical purposes (Frank et al., 2019). In an effort to meet the integration challenge through comprehensive connectivity, the number of possible digital technologies, cyber technologies, and systems connectable is uncountable and susceptible to rapid & ongoing growth (Fatorachian and Kazemi, 2018).

The three key benefits of Industry 4.0—vertical integration, horizontal integration, and end-to-end engineering—were used by other authors (Jeschke et al., 2017; Dalenogare et al., 2018) to categorize Industry 4.0 technology. CAD/CAM, integrated engineering systems, digital automation with sensors, flexible manufacturing lines, Manufacturing Execution Systems (MES), Supervisory Control and Data Acquisition (SCADA), simulation and analysis of virtual models, bid data collection and analysis, digital product systems, additive manufacturing, 3D prototyping, cloud service, and others were among the new and more consolidated technologies listed by Dalenogare et al. (2018). In addition to the technologies mentioned above, some authors (Rüßmann et al., 2015; Romero et al., 2016; Frank et al., 2019) have emphasized how Industry 4.0 depends on new autonomous and collaborative robots called collaborative robots (COBOTs) and autonomous mobile robots (AMR) that could aid workforces instead of just replacing them like in the past.

5.2.4 I4.0 AND LEAN SIX SIGMA (LSS) INTEGRATION

LSS amalgamates the Lean & Six Sigma procedure methods. Lean emphases eliminating wastage and all other parts of a procedure that don't add any value in the product, and value is demarcated as what benefits the customers. Six Sigma provides tools & approaches that aid eradicating errors & faults from the procedure. I4.0 processes can be assisted by the tools & methods used in both methods.

In addition to being a natural extension of what Lean and Six Sigma have been addressing for decades, namely that factories cannot afford waste or variability, Industry 4.0 is hailed as transformational. They claim that by combining Lean Six Sigma approaches with new technology, waste and defects will be "highlighted quickly".

The notions that LSS establishes the prerequisites for I4.0 implementation and I4.0 and LSS support one another during their separate deployments are two that frequently appear in the literature on integration.

The deployment of I4.0 technologies is made easier by LSS tools and practices since they standardize processes and cut down on human error. (Anosike et al., 2021) in addition to lowering process variability (Chiarini et al., 2021).

I4.0, also known as smart manufacturing, intelligent manufacturing & digital manufacturing, improves the flexibility, agility, and intelligence of current production processes to meet the demands of the worldwide marketplace (Zhong et al., 2017). The effects of I4.0 on Lean manufacturing & long-term organizational performance were further discussed by Kamble et al. (2019). Companies in the manufacturing sector must adapt their production, quality, and supply chain procedures to reflect the most recent developments in these fields. Recent sophisticated technologies, such as I4.0 and IoT, can be used to attain Lean goals much more quickly. On the operational performance of the organization, Khanchanapong et al. (2014) found that Lean concepts and sophisticated technology had a favorable impact. On lead-time, flexibility, cost, and quality, they found that the two manufacturing techniques significantly complemented one another. It is now necessary to find I4.0 applications in processes to integrate digital solutions in the manufacturing industry. Although it is well known that businesses have been in a learning phase for the past few decades and have implemented Lean production systems, they now need to comprehend the best practices appropriate for the current market, namely digital solutions. Now, in order to properly control Lean processes, enterprises must connect themselves with these intelligent and digital technologies.

I4.0 is distinguished by the incorporation of leading-edge technology that has transformed how industries conduct business. The sensors used in IoT are one of the key improvements. These sensors make it feasible to collect data along the whole value chain. Then, a more in-depth & compound analysis is done using those data. This more accurate data can support the Lean Six Sigma data-driven methodologies.

I4.0 technologies make it easier to achieve a high level of integration of Lean processes, which helps the business operate better overall. This new paradigm's automation and digitization aspects can aid in lowering lead times and manufacturing costs, which improves customer satisfaction. Additionally, this offers the workers the chance to learn about novel technologies that could boost their confidence. Lean manufacturing techniques might help remove obstacles to I4.0 adoption and may make it possible to start the procedure of implementing this cutting-edge technology (Kamble et al., 2019). While I4.0 delivers smartness in the plant with sophisticated information & communication methods, which provide solutions to overcome Lean implementation challenges, the Lean approach has the ability to increase productivity (Sanders et al., 2016). The use of intelligent tools in manufacturing systems can drive the whole plant setting toward mass customization. Lean automation, which blends robots and automation in Lean manufacturing, is created because of the integration of Lean & I4.0. Auto inspection and flaw detection are included in automation in the Lean manufacturing process. Manufacturing companies should investigate the advantages and potential of integrating Lean with I4.0. The goals of I4.0 & Lean manufacturing are both greater output & flexibility (Buer et al., 2018).

5.2.5 EFFECTS OF I4.0/SMART MANUFACTURING WITH LEAN INTEGRATION ON PERFORMANCE

Lean manufacturing and I4.0 have a beneficial relationship, according to the research in the literature, and difficulties implementing Lean methods can be solved by putting I4.0's recommendations into reality (Sanders et al., 2016). Modern information and communication technologies have the ability to open up new possibilities for Lean operating structures. I4.0 technologies guarantee the removal of obstacles in a number of areas relating to customers, processes, suppliers, and labor. With the applications of this novel digital & intelligent production system platform, industrial operations that have previously implemented Lean may be more modelled and controlled. I4.0 technologies have the potential to strengthen current Lean manufacturing systems for a more dependable product with benefit.

5.2.6 LEAN MANUFACTURING PROBLEMS AND I4.0'S RECOMMENDED SOLUTIONS

The correct addressing of Lean difficulties and barriers requires the synchronization of I4.0 technology with Lean concepts. Lean strives to increase a company's competitiveness while also increasing production capacity. Non-value-added items are categorized as seven wastes, or "muda," which include defects, overproduction, waiting, transportation, inventories, and over processing. Lean tools and I4.0 work together as an enabler to get rid of these wastes & increase output and efficacy.

One of the primary issues in businesses is overproduction, and this type of wastage may be avoided by keeping good order management & information flow between the shop floor and the equipment directly. This is made possible by Cyber-Physical-Space (CPS) enabled actual support data offered by I4.0 facilities (Pereira et al., 2019). In these contemporary settings, equipment gathers real-time data & responds separately based on the data composed. Real-time information made available by smart approaches allows for quicker and more effective production planning and decision-making.

Another type of waste, waiting periods combine delays with idle time for workers, machines, and other equipment due to unforeseen circumstances, which further disrupts the manufacturing process. Smart manufacturing's cyber-physical connection facility allows for precise control and monitoring of operations on the factory floor as well as the ability to make more intelligent decisions at the process or machine level to reduce waste in many ways. Integrating horizontally and vertically allows for quick input from stakeholders (Wang et al., 2016). Smart machine control eliminates unforeseen delays and interruptions in the process flow, reducing production wait times. A productive working environment is provided through smart production, maintenance, and planning systems. Machines with intelligent sensors and actuators can do autonomous maintenance for more foresighted and intelligent maintenance (Wang, 2016).

The movement of goods or resources that are not necessary for production processes is regarded as a major source of wastage in lean manufacturing. I4.0 infrastructures make it possible to choose the best path for work when processing materials on the shop floor (Wang et al., 2016).

Over processing refers to processes taken throughout the production process that provide little value. More work than is actually necessary for manufacturing is included in this waste category. Smart device-based accurate detection, diagnosis, and visualization of production processes (Posada et al., 2015) enables the elimination of redundant manufacturing stages, resulting in cost and energy savings (Shrouf et al., 2014).

Waste resulting from an excess of finished or unfinished goods or work-in-process is referred to as excess inventory (WIP). Information and communication technologies (ICT) improve value chain connectivity and communication; they provide real-time information from suppliers and customers to enable better control over excess inventory of goods and further reduce inventory costs (Posada et al., 2015). Compared to their conventional methods, autonomous Kanban and JIT build a better platform for supply chain & production optimization (Lai et al., 2019).

Uneven movements of people & machines on factory floors are known as unnecessary motion. CPSs can quickly recognize and control superfluous motion during operation and service when combined with machine learning and smart sensors (Posada et al., 2015; Jazdi, 2014). Real-time data from sensors and actuators can be used to map production streams accurately. Rework is one of the wastes that results from the poor quality of the products and calls for corrective action. I4.0 design with integrated sensors and a network can minimize faults with enhanced production process monitoring (Lee et al., 2015). Information about a machine's health and performance that is gathered through smart devices helps in defect detection (Jazdi, 2014).

5.2.7 EFFECTS OF I4.0/SMART MANUFACTURING WITH SIX SIGMA ON PERFORMANCE

Six Sigma is a statistical process grounded on "Define, Measure, Analyze, Improve, and Control" (DMAIC) methodology for measuring and analyzing processes to eliminate variability. It is a data-driven methodology that is used to reduce variances in manufacturing processes and prevent the development of errors. Big data techniques that gather accurate and useful data help to properly build Six Sigma objectives (Mendonca et al., 2018). Six Sigma improves visibility in fault finding and issue solving when assisted by digital technologies like CPS, BDAs, AI, and machine learning. Lean Six Sigma tools rely on data to drive process changes, according to Arcidiacono and Pieroni (2018), and this issue is best overcome with actual data availability, and employing data analytics techniques in the I4.0 tool set.

Because conventional statistical methods cannot handle the size of records generated by smart machines in the I4.0 environment, it is essential to provide the machines with the ability to learn on their own and forecast the future. I4.0 is equipped with cutting-edge computational methods and algorithms for data-driven real-time decisions, enabling flexible production processes. Smart sensors, robots, 3D printing, augmented and virtual reality, cloud manufacturing, and high-speed processors are examples of intelligent technologies and equipment that make businesses quick, alert, allied & well informed in their operations. The sophisticated capabilities of smart manufacturing ensure that it can handle uncertainties, maximize resource use, and create customized goods and services (Biswas, 2019). According to D'Ambrogio et al. (2008), enterprises that have begun implementing new edge technologies and using sensors, AI, cloud computing, and IoT have noted considerable gains in processing, connectivity, and data accessibility. Lean Six Sigma practitioners can assist in integrating new technologies in industrial operations if they improve their data collecting abilities and create insights that can be put to use.

Because client needs and tastes are constantly changing, designing manufacturing processes so that businesses can produce customized products with exact levels in the shortest amount of time presents significant challenges to strategic decision-makers (Sodhi, 2020). To deal with real data, this movement in consumer behavior needs the backing of cutting-edge digital and communication technology. To meet client needs, businesses must plan to keep good relationships and networks with qualified suppliers and partners. While Six Sigma focuses on standardization by reducing process variation, Lean enables waste detection and elimination.

Advanced data approaches are vital for making efficient judgements for quality issues because traditional data analysis methods have the disadvantage of taking more time and money. Modern approaches such as machine learning, decision trees, text mining, video mining, and AI can be used at each stage of LSS to produce more certain and predictable decisions.

5.2.8 I4.0's Sustainability Features

I4.0 design principles and technologies have the potential to assist sustainability in terms of the economy, environment, and society (Stock and Seliger, 2016). One of the important technologies, additive manufacturing, promotes the circular economy, and many technical and communication technologies help to increase the workforce's knowledge and skill set. The sustainability strategy, along with I4.0 principles and technologies, helps to achieve resource efficiency, manufacturing flexibility, and lower energy use (Morteza, 2019). I4.0's smart data techniques provide high-quality products at an affordable price with less adverse environmental effects.

5.3 CONCLUSIONS AND FUTURE SCOPE

The interaction between Lean, Six Sigma, and I4.0 is elaborated in the present chapter. This study investigates how I4.0 technologies affect the Lean and Six Sigma production system philosophies. The report also discusses how new technologies work with manufacturing I4.0's sustainability features. The principles and technologies underlying this new manufacturing shift have favorable effects on the social, economic, and environmental pillars of sustainability. I4.0 opens-up a new route for creating specialized goods and services in response to continuously shifting consumer demands and preferences. CPSs, IoT, data science, AI, and other I4.0 assets can be used to design an optimal and optimized process flow in production systems.

I4.0's intelligent and perceptive technologies help people make important decisions and provide significant feedback at every stage of the LSS approach. ICT-based solutions can better solve problems with Lean implementation. Through the real-time and accurate information sharing mechanisms provided by digital technology, Lean pillars such as JIT, automation (Jidoka), Kanban, TPM, and VSM become more dependable. Increasing operational efficiency and offering better products and services in accordance with client feedback are two additional benefits of integrating these manufacturing processes. Lean Six Sigma practitioners need to get ready for the exposure to these new technologies with the arrival of advanced technology. Every company should invest in adopting I4.0 technology in their manufacturing facilities at this time of transition. Utilizing these technologies will enable the ability to identify and address issues that are present on the shop floor of businesses committed to sustainable growth. The goal of the current study was to demonstrate the value of combining Lean and Six Sigma with I4.0 and the sustainability aspects of this new paradigm of advanced technologies. However, there is still room to validate these findings using the conceptual model in a real-world organization to uncover additional synergy & interaction between these methodologies.

REFERENCES

Agrifoglio, R., Cannavale, C., Laurenza, E. and Metallo, C. (2017). "How emerging digital technologies affect operations management through co-creation. Empirical evidence from the maritime industry". Production Planning & Control, Vol. 28 (16), 1298–1306.

Al-Aomar, R.A. (2011). "Applying 5S LEAN Technology: An infrastructure for continuous process improvement". International Journal of Industrial and Manufacturing Engineering, Vol. 5 (12), 2645–2650.

Anosike, A., Alafropatis, K., Garza-Reyes, J.A., Kumar, A., Luthra, S. and Rocha-Lona, L. (2021). "Lean manufacturing and internet of things—A synergetic or antagonist relationship?". Comput. Ind. Vol. 129, 103464.

Arcidiacono, G. and Pieroni, A. (2018). "The revolution of Lean Six Sigma 4.0". International Journal on Advanced Science Engineering Information Technology, Vol. 8, No. 1, pp.141–149.

Bakator, M., Dordevic, D., Vorkapic, M. and Ceha, M. (2019). "Modelling the use of Industry 4.0 technologies with Lean manufacturing". Proceeding of the International Symposium Engineering Management and Competitiveness, Zrenjanin, Serbia, 21–22 June, pp.41–16.

Baudin, M. (2007). "Working with machines: the nuts and bolts of lean operations with Jidoka". Cambridge, MA: CRC Press.

Bhamu, J. and Sangwan, K.S. (2014). "Reduction of post-kiln rejections for improving sustainability in ceramic industry: a case study". Procedia CIRP, Vol. 26, pp.618–623.

Bibby, L., and Dehe, B. (2018). "Defining and assessing industry 4.0 maturity levels-case of defence sector". Production Planning & Control Vol. 29 (12), 1030–1043.

Biswas, S. (2019). "Implications of Industry 4.0 vis-à-vis Lean Six Sigma: a multi-criteria group decision approach". Proceedings of J.D. Birla International Management Conference on 'Strategic Management in Industry 4.0, Kolkata, India, pp.1–14.

Buer, S.V., Strandhagen, J.O. and Chan, F.T.S. (2018). "The link between Industry 4.0 and Lean manufacturing: mapping current research and establishing a research agenda". International Journal of Production Research, Vol. 56 (8), pp.2924–2940.

Cheng, L.J. (2018). "Implementing Six Sigma within Kaizen events, the experience of AIDC in Taiwan". The TQM Journal Vol. 30 (1), 43–53.

Chiarini, A.; Kumar, M. (2021). "Lean Six Sigma and Industry 4.0 integration for Operational Excellence: Evidence from Italian manufacturing companies". Prod. Plan. Control, Vol. 32, 1084–1101.

D'Ambrogio, A., Gianni, D., Iazeolla, G. and Pieroni, A. (2008). "Distributed simulation of complex systems by use of an HLA-transparent simulation language". In ICSC 2008, Asia Simulation Conference-7th International Conference on System Simulation and Scientific Computing, pp.460–467.

Dalenogare, L.S., Benitez, G.B., Ayala, N.F. and Frank, A.G. (2018). "The expected contribution of Industry 4.0 technologies for industrial performance". International Journal of Production Economics Vol. 204, 383–394.

Dutta, S. and Jaipuria, S. (2020). "Reducing packaging material defects in beverage production line using Six Sigma methodology". International Journal of Six Sigma and Competitive Advantage, Vol. 12 (1), pp.59–82.

Fatorachian, H. and Kazemi, H. (2018). "A critical investigation of Industry 4.0 in manufacturing: theoretical operationalization framework". Production Planning & Control, Vol. 29 (8), 633–644.

Frank, A.G., Dalenogare, L.S., and Ayala., N.F. (2019). "Industry 4.0 technologies: Implementation patterns in manufacturing". International Journal of Production Economics, Vol. 210, 15–26.

Ghobakhloo, M. (2018). "The future of manufacturing industry: A strategic roadmap toward Industry 4.0". Journal of Manufacturing Technology Management Vol. 29 (6), 910–936.

Jazdi, N. (2014). "Cyber physical systems in the context of Industry 4.0".In AQTR 2014, Proceedings of IEEE International Conference on Automation, Quality and Testing, Robotics, pp.1–4.

Jeschke, S., Brecher, C., Meisen, T., Özdemir, D., and Eschert, T. (2017). Industrial Internet of Things and Cyber Manufacturing Systems. Berlin: Springer.

Kamble, S., Gunasekaran, A. and Dhone, N.C. (2019). "Industry 4.0 and Lean manufacturing practices for sustainable organizational performance in Indian manufacturing companies". International Journal of Production Research, Vol. 58 (5), pp.1–19.

Karnjanasomwong, J. and Thawesaengskulthai, N. (2019). "Dynamic sigma-TRIZ solution model for manufacturing improvement and innovation, case study in Thailand". International Journal Six Sigma and Competitive Advantage, Vol. 11 (2/3), pp.114–156.

Khadem, M., Ali, S.A. and Seifoddini, H. (2008). "Efficacy of lean metrics in evaluating the performance of manufacturing systems". International Journal of Industrial Engineering, Vol. 15 (2), 176–184.

Khanchanapong, T., Prajogo, D., Sohal, A.S., Cooper, B.K., Yeung, A.C.L. and Cheng, T.C.E. (2014). "The unique and complementary effects of manufacturing technologies and Lean practices on manufacturing operational performance". International Journal of Production Economics, Vol. 153, pp.191–203.

Kolberg, D., Knobloch, J., and Zuhlke, D. (2017). "Towards a lean automation interface for Workstations". International Journal of Production Research Vol. 55 (10), 2845–2856.

Lai, N.Y.G., Wong, K.H., Halim, D., Lu, J. and Kang, H.S. (2019). "Industry 4.0 enhanced Lean manufacturing". Proceedings of Eighth International Conference on Industrial Technology and Management (ICITM), Cambridge, UK, pp.206–211.

Lee, J., Bagheri, B. and Kao, H.A. (2015). "A cyber-physical systems architecture for Industry 4.0-based manufacturing systems". Manufacturing Letters, Vol. 3, pp.18–23.

Lu, Y. (2017). "Industry 4.0: A survey on technologies, applications and open research Issues". Journal of Industrial Information Integration Vol. 6 (2), 1–10.

Matzka, J., Di Mascolo, M. and Furmans, K. (2012). "Buffer sizing of a Heijunka Kanban System". Journal of Intelligent Manufacturing Vol. 23 (1), 49–60.

Mendonca Jr., F., Montenegro, M., Thadani, R., Pedroso, G.A.C. and de Oliveira, M.A. (2018) "Industry 4.0 as a way to enhance Lean manufacturing and Six Sigma". Proceeding of the Fifth European Lean Educator Conference, pp.152–160.

Monden, Y. (2011). Toyota production system: an integrated approach to just in time. New York: Productivity Press.

Morteza, G. (2019). "Determinants of information and digital technology implementation for smart manufacturing". International Journal of Production Research, Vol. 58 (8), pp.1–22.

Ohno, T. (1988). Toyota Production System: beyond large-scale production. New York: CRC Press.

Pereira, C.A., Dinis-Carvalho, J., Alves, C.A. and Arezes, P.M. (2019). "How Industry 4.0 can enhance Lean practices". FME Transactions, Vol. 47 (4), pp.810–822.

Posada, J., Toro, C., Barandiaran, I., Oyarzun, D., Stricker, D. and De Amicis, R. (2015). "Visual computing as a key enabling technology for Industries 4.0 and industrial internet". IEEE Computer Graphics and Applications, Vol. 35 (2), pp.26–40.

Powell, D., Romero, D., Gaiardelli, P., Cimini, C. and Cavalieri, S. (2018). Towards digital lean cyber-physical production systems: Industry 4.0 technologies as enablers of leaner production. In Advances in Production Management Systems. Smart Manufacturing for Industry 4.0: IFIP WG 5.7 International Conference, APMS 2018, Seoul, Korea, August 26–30, 2018, Proceedings, Part II, 353–362. Springer International Publishing.

Romero, D., Stahre, J., Wuest, T., Noran, O., Bernus, P., Fast-Berglund, Å. and Gorecky, D. (2016). "Towards an operator 4.0 typology: a human-centric perspective on the fourth industrial revolution technologies". In Proceedings of the International Conference on Computers & Industrial Engineering – CIE46.

Rüßmann, M., Lorenz, M., Gerbert, P., Waldner, M., Justus, J., Engel, P. and Harnisch, M. (2015). "Industry 4.0: The future of productivity and growth in manufacturing industries". Boston Consulting Group.

Salah, S., Rahim, A. and Carretero, J.A. (2010). "The integration of Six Sigma and lean Management". International Journal of Lean Six Sigma Vol. 1 (3), 249–274.

Salum, L. (2000). "The cellular manufacturing layout problem". International Journal of Production Research Vol. 38 (5), 1053–1069.

Sanders, A., Elangeswaran, C. and Wulfsberg, J. (2016). "Industry 4.0 implies Lean manufacturing: research activities in Industry 4.0 function as enablers for Lean manufacturing". Journal of Industrial Engineering and Management, Vol. 9 (3), pp.811–833.

Shahin, M., Chen, F.F., Bouzary, H. and Krishnaiyer, K. (2020). "Integration of Lean practices and Industry 4.0 technologies: smart manufacturing for next-generation enterprises". The International Journal of Advanced Manufacturing Technology, Vol. 107 (5–6) pp.2927–2936.

Shingo, S. (1996). Quick changeover for operators: the SMED system. Cambridge, MA: Productivity press.

Shrouf, F., Ordieres, J. and Miragliotta, G. (2014). "Smart factories in Industry 4.0: a review of the concept and of energy management approaches in production based on the internet of things Paradigm". In IEEE International Conference on Industrial Engineering and Engineering Management, pp.697–701.

Snee, R.D. (2010). "Lean Six Sigma–getting better all the time". International Journal of Lean Six Sigma Vol. 1 (1), 9–29.

Sodhi, H. (2020). "When Industry 4.0 meets Lean Six Sigma: a review". Industrial Engineering Journal, Vol. 13 (1), pp.1–12.

Stock, T. and Seliger, G. (2016). "Opportunities of sustainable manufacturing in Industry 4.0". Procedia CIRP, Vol. 40, pp.536–541.

Vaidya, S., Ambad, P. and Bhosle, S. (2018). "Industry 4.0 – a glimpse". Procedia Manufacturing, Vol. 20, pp.233–238.

Wang, S., Wan, J., Li, D. and Zhang, C. (2016). "Implementing smart factory of Industry 4.0: an Outlook". International Journal of Distributed Sensor Networks, Vol. 12 (1) pp.1–10.

Zhong, R.Y., Xu, X., Klotz, E. and Newman, S.T. (2017). "Intelligent manufacturing in the context of Industry 4.0: a review". Engineering, Vol. 3 (5) pp.616–630.

6 Exploring Critical Success Factors for LSS 4.0 Implementation

A Combined Systematic Literature Review and Interpretive Structural Modelling Approach

*Dounia Skalli, Abdelkabir Charkaoui,
Anass Cherrafi, and Jose Arturo Garza-Reyes*

6.1 INTRODUCTION

Today we are experiencing a shift in business and market dynamics that goes beyond globalization and increased competition (Cherrafi et al., 2016b; Garza-Reyes, 2015; Skalli et al., 2023; Sony et al., 2021) to encompass the new era of digitization, stakeholder empowerment and changing customer orientations (Anass et al., 2021a; Buer et al., 2021). As digital awareness grows among customers and intense competition puts pressure on, manufacturers must rethink their operating methods and models to accommodate the advanced technologies offered by Industry 4.0. These novel technologies have become a mechanism to gain competitive advantage, and improve operations performance (Moeuf et al., 2018).

Classically, efficiency, productivity, profitability, and customer satisfaction have been the primary concerns of organizations. Actually all organizations face challenges related to resilience, viability, connectivity and sustainability (Alqudah et al., 2020; Ghobakhloo, 2020). In response to the increasing challenges, organizations have already begun to move to smart and efficient business operations. However, since the advent of the Fourth Industrial Revolution, the concept of Industry 4.0 has been adopted by manufacturers across value chain networks and production systems, through the digitization of the entire process structure to analyze data more efficiently (Tortorella et al., 2019). Organizations have been forced to change the way they work

DOI: 10.1201/9781003381600-6

and are looking for a high level of operational excellence and new organizational capabilities. Various strategies, such as Six Sigma, Lean, green, agile and resilience are used by manufacturing companies to improve efficiency and performance (Ali et al., 2021). Numerous studies have confirmed that Lean Six Sigma can be considered as an approach to improve operational excellence (Buer et al., 2021). In recent years, practitioners have increasingly focused on integrating advanced technologies with continuous improvement methods and redesigning their processes to capitalize on the benefits granted by Industry 4.0 technologies to improve the operational performance (Lameijer et al., 2021). Currently, most organizations are looking for their way to LSS 4.0 but still facing difficulties due to some challenges and barriers (Sony & Naik, 2020). Other authors (Anass et al., 2021a; Antony et al., 2022; Lameijer et al., 2021; Skalli et al., 2023) reported in their studies that the combination of Lean Six Sigma with Industry 4.0 technologies is an important organizational philosophy, which plays a significant role in promoting quality and efficiency, enabling the reduction of waste, losses and process variability to achieve operational excellence. The effective implementation of LSS 4.0 depends on several factors that play a critical role in achieving a successful integrated approach. It is therefore necessary to identify and evaluate the critical success factors (CSF) for its implementation. These critical success factors are used to guide managers on their journey to LSS 4.0 by understanding the critical factors for achieving productivity and efficiency. In the implementation phase of LSS 4.0 in the manufacturing industry, it is recommended to know the success and failure factors. This study is one of the early studies to examine the mutual relationship between the Critical Success Factors.

This paper aims to identify and classify relevant CSFs to the adoption of Lean Six Sigma (LSS) and Industry 4.0 (I4.0) as an integrated approach using a two-phase study design.

The main objectives of our study are:

1. To identify the CSFs that can help companies successfully implement LSS 4.0 using a literature review.
2. To contextualize the relationship between them using the Interpretive Structural Modeling (ISM) method.
3. To generate a model to understand the keys components that will help companies towards the effective implementation of LSS 4.0.

The rest of this paper is organized as follows: Section 2 describes a review of the various CSFs identified in the literature. Section 3 explains the research methodology while Section 4 gives an analysis of the CSFs through the ISM technique. Finally, the results, discussion and implications are presented in Section 5.

6.2 SYSTEMATIC LITERATURE REVIEW OF LSS 4.0 CRITICAL SUCCESS FACTORS

This section presents a summary of the literature on Lean, Six Sigma and Industry 4.0.

6.2.1   BACKGROUND ON LSS4.0 AS AN INTEGRATED APPROACH

Before addressing the research questions, a brief overview of LSS and I4.0 applications in the manufacturing sector is required.

Lean Six Sigma is the combination of Lean and Six Sigma, the two best used continuous improvement methodologies in manufacturing, recognized for their effectiveness in increasing business performance (Albliwi et al., 2015). Lean and Six Sigma were launched and developed in the manufacturing environment to help operations managers reduce waste, simplify the production line and improve quality (Drohomeretski et al., 2014). They are now widely perceived as one of the most effective ways to increase an organization's competitiveness (Shah et al., 2008). Six Sigma is a business improvement strategy that aims to identify and eliminate the causes of process defects by focusing on those that are relevant to customers (Snee, 2010). Combining Lean Six Sigma with other strategies such as green, resilience, circular economy, and, more recently, Industry 4.0, can further improve business performance (Skalli, Charkaoui, & Cherrafi, 2022). With the advent of the fourth industrial revolution called 'Industry 4.0', Lean Six Sigma 4.0 has emerged as an initiative to improve operational excellence by combining digital technologies with Lean and Six Sigma tools. However, studies on the factors affecting LSS 4.0 implementation are limited. The literature review reveals that the CSFs for Lean Six Sigma and Industry 4.0 were reviewed separately, because no previous research related to the field was found. For this reason, input from industry and academic experts allowed us to refine the CSFs to fit LSS 4.0.

Recently, the relationship between Lean Six Sigma and Industry 4.0 has been well examined (Antony et al., 2022; Buer et al., 2021; Kumar et al., 2016; Skalli, Charkaoui, & Anass, 2022; Tortorella et al., 2019). Researchers suggest that LSS and I4.0 are synergetic. LSS 4.0 will assist managers to handle complex issues related to product quality and customer satisfaction easily by making a proper integration between LSS tools and advanced technologies (Buer et al., 2021). Therefore, it is necessary to create a strategic framework for its implementation by identifying and analyzing the key success factors. This knowledge can assist in the implementation of LSS 4.0.

6.2.2   IDENTIFICATION OF CSFs FOR ADOPTING LSS 4.0

A CSF is an enabling or helping element that moves things forward or makes it easier for someone to achieve something. LSS 4.0 is guided by a number of theoretical elements, including drivers, barriers, challenges and CSFs. Therefore, these elements must be identified and understood for the most successful deployment of LSS 4.0. In this study, the critical success factors describe the vital components that must be taken into consideration by an organization on its way to LSS 4.0. In our study, CSFs were identified using a systematic literature review (SLR) following the five sequential phases suggested by (Denyer & Tranfield, 2009) as described in Figure 6.1.

Initially, an extended search, based on numerous combinations of the set of keywords as described in Table 6.1, in the following databases (Scopus, Springer, Emerald, Taylor & Francis, IEEE, Google Scholar, Elsevier), and using the Boolean

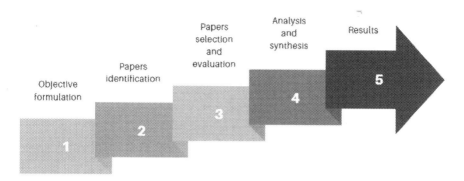

FIGURE 6.1 Systematic literature review approach (SLR).

TABLE 6.1
The Systematic Literature Review Overview

Study type	Qualitative
Keywords	Six sigma, Lean, Lean six sigma, Lean 4.0, LSS4.0, Industry 4.0, CSFs using Boolean operators (i.e. AND and OR)
Databases	Scopus, Springer, Emerald, Taylor & Francis, IEEE, Google Scholar, Elsevier
Research period	From 2011 to December 2022
Inclusion criteria	Indexed scientific papers written in English:
	(1) journals; (2) conference papers, and (3) books and chapters of books.
Exclusion criteria	White papers
	Non-peer-reviewed articles
	Articles not relevant to the research objectives
Software tools	Zotero
	Nvivo

connectors 'OR' and 'AND', was performed to allow a rigorous and advanced identification and selection of articles. After duplicate removal, in the second step, the articles were selected using inclusion and exclusion criteria previously defined and presented in Table 6.1. We used the Context-Intervention-Mechanism-Outcome (CIMO) approach suggested by (Briner & Denyer, 2012) to facilitate the exclusion and inclusion criteria. As a result, we included only peer-reviewed journal articles; conference papers and book chapters written in English, while white papers and non-peer-reviewed articles were excluded. First 72 articles were retrieved. After applying the inclusion and exclusion criteria, 52 articles were finally selected for analysis. We adopted a SLR approach coupled with thematic analysis for rigorous field analysis, to identify the most relevant CFSs discussed by scholars. We used thematic synthesis as it is considered the most structured, appropriate, and efficient method to generate themes (Braun & Clarke, 2006) with the help of the NVivo software. The thematic analysis resulted in ten different factors being identified.

TABLE 6.2
List of LSS 4.0 CSFs

Critical Success Factors	I4.0	LSS	References
Aligning Industry 4.0 and LSS initiatives with organizational strategy and vision statements (CSF1)	X	X	(Cherrafi et al., 2016; Moeuf et al., 2018; Sony & Naik, 2020)
Top management commitment and Leadership (CSF2)	X	X	(Belhadi et al., 2019; Cherrafi, Elfezazi, Chiarini, et al., 2017; Ciano et al., 2019; Javaid & Haleem, 2020; Kumar, 2007; Sony et al., 2021; Sony & Naik, 2020; Yadav et al., 2020)
Employees' awareness, training and involvement (CSF3)	X	X	(Sony & Naik, 2020)
Allocation of resources and infrastructure (CSF4)	X	X	(Antony et al., 2022; Pozzi et al., 2021; Sony & Naik, 2020; Yadav et al., 2020)
Clear objectives, goals and responsibilities (CSF5)	X	X	(Cherrafi, Elfezazi, Chiarini, et al., 2017; Ghobakhloo, 2020; Laureani & Antony, 2012; Pozzi et al., 2021)
Appropriate workforce skills and managers' expertise (CSF6)	X	X	(Belhadi et al., 2019; Cherrafi, Elfezazi, Chiarini, et al., 2017; Javaid & Haleem, 2020; Kumar, 2007; Lameijer et al., 2021)
Promote knowledge, motivation, communication and Change management (CSF7)	X	X	(Ciano et al., 2019; Lameijer et al., 2021; Pozzi et al., 2021) (Antony et al., 2021a)
Suitable selection and prioritization of projects and Technologies (CSF8)	X	X	(Antony et al., 2022; Pozzi et al., 2021; Sony et al., 2021; Sony & Naik, 2020; Yadav et al., 2020)
Cyber security management (CSF9)	X		(Sony et al., 2021)
Effective readiness for LSS and I4.0 (CSF10)	X	X	(Sony et al., 2021)

CSFs were identified by analyzing the existing literature and then including contributions from ten experts with knowledge on LSS and Industry 4.0 and more than ten years of experience. The list of selected CSFs was further discussed and refined with a panel of experts and presented in Table 6.2.

6.3   RESEARCH METHODOLOGY

The main objective of our study is to find the relationships between the CSFs identified in the literature and thus determine their dominance in the implementation of LSS 4.0 initiatives. To achieve our research objective, we used a three-step research methodology, first we started with an extensive literature review to identify the most discussed factors within the researchers, then we provided insights from ten professional experts to classify and determine the variables, finally we used the ISM approach, a well- established interactive technique to identify and analyze contextual relationships among selected LSS 4.0 CSFs.

There are multiple techniques for multi-criteria decision making. The choice of technique depends on the efficacy, applicability, relevance, and the form of outcome preferred by the candidates. Interpretive structural modeling (ISM), analytic network process (ANP), and analytic hierarchy process (AHP) are the most well-known multi-criteria decision making (MCDM) techniques (Drohomeretski et al., 2014).

The ISM method with MICMAC analysis has been used in several studies (Ali et al., 2020; Cherrafi, Elfezazi, Garza-Reyes, et al., 2017; Luthra et al., 2020; Yadav & Desai, 2017). The ISM technique is a well-known method first introduced by Warfield (1974) and Sage (1977). It has been used by researchers to understand and analyze the interactions between different problem variables in several industries.

ISM is based on using the practical expertise and know-how of the experts to solve a complex problem or area of study regarding the relationships between its variables. The idea is to break down the problem into several sub-systems to build a multi-level structural model.

To develop our ISM model, we proceeded with the following steps (see Figure 6.2):

Step 1: Identification of CSFs based on a literature review (Fifteen CSFs).
Step 2: Determination of the variables that are the defined CSFs (CSF 1 to CSF 10) through expert opinions.
Step 3: Contextualization of the relationships between the CSFs.
Step 4: Development of Structural Self-Interaction Matrix (SSIM) to demonstrate the pairwise relations between the CSFs.
Step 5: Development of an initial reachability matrix from the SSIM.

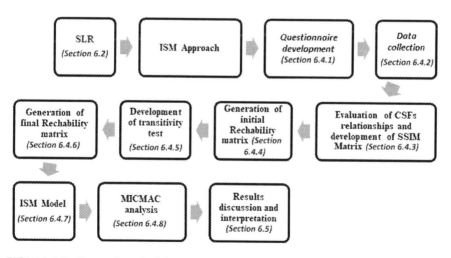

FIGURE 6.2   Research methodology.

Step 6: Application of a transitivity test to establish the final reachability matrix.
Step 7: Development of level partitions on the basis of the reachability, antecedent and intersection sets of each factor.
Step 8: Generation of directed graph converted to the ISM model.

6.4 INTERPRETIVE STRUCTURAL MODELING – ISM MODEL

6.4.1 QUESTIONNAIRE DEVELOPMENT

To conduct the analysis of the ten critical success factors for the effective implementation of LSS 4.0 previously identified in section 2 (see Table 6.1), 18 manufacturing experts from the automotive sector and six academic researchers were approached directly through the authors' network and via Linkedin. Meetings or phone calls were held to explain the objectives of this study and to convince them to participate. As a result, ten professional experts, from different sectors, positions and experience agreed to participate in the study, including seven industrial experts and three senior researchers. All participants had at least ten years of experience in conducting improvement projects. Table 6.3 summarizes the experts' profile. All industrial participants belonged to companies that had successfully completed a LSS 4.0 pilot project or were in the process of initiating one.

6.4.2 DATA COLLECTION

The ISM technique is based on using the experts' opinions to identify the contextual relationships between the identified variables. For that, two meetings with the experts were therefore organized. The first meeting was devoted to quantifying the pertinence and importance of the CSFs identified by the literature review (LR) on a scale of High-Moderate-Low reported in Table 6.4, while the second meeting allowed the experts to establish the relationships in the form of "factor 1 affects factor 2", according to (Kumar et al. 2016).

TABLE 6.3
The Experts' Profile

Position	Number	Years of Experience
IT Manager	1	11
Quality Manager	1	12
General Director	1	19
Plant Manager	1	15
Operations Manager	1	15
Kaizen Manager	1	12
Site Manager	1	20
University Professors	3	20>

TABLE 6.4
Importance of CSFs in Implementing LSS4.0

SN	Critical Success Factor	Importance
1	Aligning Industry 4.0 and LSS initiatives with organizational strategy and vision statements (CSF1)	High
2	Top management commitment and leadership (CSF2)	High
3	Employees' awareness, training and involvement (CSF3)	High
4	Allocation of resources and infrastructure (CSF4)	High
5	Clear objectives, goals and responsibilities (CSF5)	High
6	Appropriate workforce skills and managers' expertise (CSF6)	High
7	Promote knowledge, motivation, communication and change management (CSF7)	Moderate
8	Suitable selection and prioritization of projects and technologies (CSF8)	High
9	Cyber security management (CSF9)	High
10	Effective readiness for LSS and I4.0 (CSF10)	High

6.4.3 SSIM Development

To define the interaction between two CSFs (row i and column j) and the sense of this relationship, a group of experts from both industry and academia that were familiar with LSS and I4.0 were interviewed. The following four symbols were used to indicate the associated direction of the relationship between two CSFs.

V – CSF i will affect CSF j
A – CSF j will affect CSF i
X – CSF i and j affect each other.
O – CSF i and j are not associated

Based on the contextual relationships developed between the CSFs, considering the expert feedback, the SSIM is developed using the V, A, X, and O symbols as shown in Table 6.5.

6.4.4 Initial Reachability Matrix Formation

The reachability matrix is obtained by converting the symbols V, A, X and O used in SSIM into binary values 1 and 0. We proceeded as follows:

- If the symbol (i, j) is V, it is assigned 1 in the initial reachability matrix and (j, i) receives 0.
- If the symbol (i, j) in SSIM is A, it is assigned 0 in the initial reachability matrix and (j, i) becomes 1.
- If the symbol (i, j) in SSIM is X, then we replace (i, j) and (j, i) by 1.
- If the symbol (i, j) in SSIM is O, we substitute (i, j) and (j, i) by 0.

TABLE 6.5
SSIM of CSFs

CSFs	CSF 10	CSF 9	CSF 8	CSF 7	CSF 6	CSF 5	CSF 4	CSF 3	CSF 2	CSF 1
CSF1	O	X	O	O	A	X	O	A	V	-
CSF 2	O	A	X	O	X	V	A	O	-	-
CSF 3	V	O	O	V	A	V	X	-	-	-
CSF 4	V	O	V	V	O	V	-	-	-	-
CSF 5	O	X	A	V	A	-	-	-	-	-
CSF 6	O	X	V	O	-	-	-	-	-	-
CSF 7	V	O	O	-	-	-	-	-	-	-
CSF 8	O	X	-	-	-	-	-	-	-	-
CSF 9	O	-	-	-	-	-	-	-	-	-
CSF10	-	-	-	-	-	-	-	-	-	-

TABLE 6.6
Initial Reachability Matrix (IRM)

CSF	CSF 1	CSF 2	CSF 3	CSF 4	CSF 5	CSF 6	CSF 7	CSF 8	CSF 9	CSF 10
CSF1	1	1	0	0	1	0	0	0	1	0
CSF 2	0	1	0	0	1	1	0	1	0	0
CSF 3	1	0	1	1	1	0	1	0	0	1
CSF 4	0	1	1	1	1	0	1	1	0	1
CSF 5	1	0	0	0	1	0	1	0	1	0
CSF 6	1	1	1	0	1	1	0	1	1	0
CSF 7	0	0	0	0	0	0	1	0	0	1
CSF 8	0	1	0	0	1	0	0	1	1	0
CSF 9	1	1	0	0	1	1	0	1	1	0
CSF10	0	0	0	0	0	0	0	0	0	1

Having applied the preceding rules, the initial accessibility matrix obtained is presented in Table 6.6.

6.4.5 FINAL REACHABILITY MATRIX FORMATION

We applied transitivity in the initial reachability matrix which states that when variable "A" is related to variable "B" and variable "B" is linked to variable "C", then variable "A" will necessarily be related to variable "C". The final reachability matrix is shown in Table 6.7.

6.4.6 LEVEL PARTITIONS

Reachability and antecedent sets of each CSF were derived from the final reachability matrix as shown in Table 6.6. The reachability set includes the CSF itself and the

TABLE 6.7
Final Reachability Matrix (FRM)

CSF	CSF 1	CSF 2	CSF 3	CSF 4	CSF 5	CSF 6	CSF 7	CSF 8	CSF 9	CSF 10
CSF1	1	1	0	0	1	1*	1*	1*	1	0
CSF 2	1*	1	1*	0	1	1	1*	1	1*	0
CSF 3	1	1*	1	1	1	0	1	1*	1*	1
CSF 4	1*	1	1	1	1	1*	1	1	1*	1
CSF 5	1	1*	0	0	1	1*	1	1*	1	1*
CSF 6	1	1	1	1*	1	1	1*	1	1	1*
CSF 7	0	0	0	0	0	0	1	0	0	1
CSF 8	1*	1	0	0	1	1*	1*	1	1	0
CSF 9	1	1	1*	0	1	1	1*	1	1	0
CSF10	0	0	0	0	0	0	0	0	0	1

TABLE 6.8
Level Partitions

CSFs	Antecedent Set	Reachability Set	Intersection Set	Level
CSF 1	1,2,5,6,7,8,9	1,2,3,4,5,6,8,9	1,2,5,6,8	4
CSF 2	1,2,3,5,6,7,8,9	1,2,3,4,5,6,8,9	1,2,3,5,6,8,9	3
CSF 3	1,2,3,4,5,7,8,9,10	2,3,4,6,9	2,3,4,9	5
CSF 4	1,2,3,4,5,6,7,8,9,10	3,4,6	3,4,6	5
CSF 5	1,2,5,6,7,8,9,10	1,2,3,4,5,6,8,9	1,2,5,6,8,9	3
CSF 6	1,2,3,4,5,6,7,8,9,10	1,2,4,5,6,8,9	1,2,4,5,6,8,9	6
CSF 7	7,10	1,2,3,4,5,6,7,8,9,	7	2
CSF 8	1,2,5,6,7,8,9	1,2,3,4,5,6,8,9	1,2,5,6,8,9	3
CSF 9	1,2,3,5,6,7,8,9	1,2,3,4,5,6,8,9	1,2,3,5,6,8,9	3
CSF 10	10	3,4,5,6,7,10	10	1

others that were driven by it and the antecedent set consists of the CSF itself and the others on which it depends. The intersection set for each CSF was derived from reachability and antecedent set. CSFs whose reachability and intersection set are the same have been placed at the first level of the model, meaning that they will not help drive any other CSF. After recognizing the first level, it was then removed from all sets and the same process is applied to find the other levels of the models. The identified levels are presented in Table 6.8, which is the basis for the ISM model construction. From Table 6.8, it is clear that "Effective readiness for LSS and I4.0" is identified at the top level, which is Level 1.

6.4.7 ISM-Based Model

Based on the final reachability matrix, the structural model of the CSFs is generated. As shown in Figure 6.3, six levels of CSFs are derived from the level partition iteration process and are represented in the ISM model diagraph. The 'Appropriate workforce skills and managers' expertise' is a highly significant CSF to LSS 4.0 implementation as it represents the bottom line of the ISM hierarchy. 'Employees awareness, training and involvement' and 'Allocation of resources and infrastructure' may be influenced by 'Appropriate workforce skills and managers' expertise' as evident from the ISM-based model proposed in Figure 6.3. The 'Appropriate workforce skills and managers' expertise' CSF may further lead to 'Aligning Industry 4.0 and LSS initiatives with organizational strategy and vision statements'. The next four CSFs in Level Four 'Cyber security management', 'Top management commitment and Leadership', 'Clear objectives, goals, and responsibilities', and 'Suitable selection and prioritization of projects and Technologies' are all directly correlated to 'Promoting knowledge, motivation, communication and Change management'

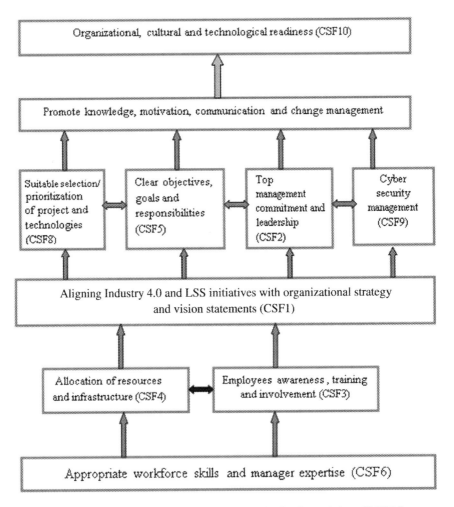

FIGURE 6.3 ISM-based model for CSFs to the effective implementation of LSS4.0.

which leads to 'Organizational, cultural, and technological readiness' which has been analyzed in the first position.

6.4.8 MICMAC Analysis

The main goal of the MICMAC analysis is to study the driving power and the dependence power of variables (Ravi & Shankar, 2005). Hence, based on the driving and dependence power calculated in Table 6.9 we generated the MICMAC analysis as shown in Figure 6.3. According to (Kumar et al., 2016), the driving power of the variable is the total of the corresponding row in the final reachability matrix and the dependency power of the CSF is the total of the corresponding column in the final reachability matrix. The dependence power of a variable indicates whether the CSF depends on other CSFs, similarly, the driving power shows whether the CSF influences other CSFs. Referring to Table 6.7, CSF1 gets a sum of 7 in the row; this means that CSF1 influences six other CSFs. Similarly, CSF1 gets a sum of 8, in the column, which means that CSF is influenced by seven other CSFs. The MICMAC analysis clusters the CSFs into four different groups, as follows:

1. Autonomous CSFs: The CSF variables under this cluster had weak driving and weak dependence power. As shown in Figure 6.4, these CSFs were represented in Cluster I and no CSFs fall in this category in our study.
2. Dependent CSFs: In cluster II, we classified CSFs having a weak driving power but a strong dependence power. Two variables, 'Promote knowledge, motivation, communication and Change management' (CSF7) and 'Effective readiness for LSS and I4.0' (CSF 10) are included in this group.
3. Linkage CSFs: These CSFs had a strong driving power as well as a strong dependence power. These CSFs were regrouped in cluster III. There are six

TABLE 6.9
Driving Power and Dependence Power

CSF	CSF 1	CSF 2	CSF 3	CSF 4	CSF 5	CSF 6	CSF 7	CSF 8	CSF 9	CSF 10	Driving Power
CSF1	1	1	0	0	1	1*	1*	1*	1	0	7
CSF 2	1*	1	1*	0	1	1	1*	1	1*	0	8
CSF 3	1	1*	1	1	1	0	1	1*	1*	1	9
CSF 4	1*	1	1	1	1	1*	1	1	1*	1	10
CSF 5	1	1*	0	0	1	1*	1	1*	1	1*	8
CSF 6	1	1	1	1*	1	1	1*	1	1	1*	10
CSF 7	0	0	0	0	0	0	1	0	0	1	2
CSF 8	1*	1	0	0	1	1*	1*	1	1	0	7
CSF 9	1	1	1*	0	1	1	1*	1	1	0	8
CSF10	0	0	0	0	0	0	0	0	0	1	1
Dependence power	8	8	5	3	8	7	9	8	8	6	

CSFs in this cluster, 'Aligning Industry 4.0 and LSS initiatives with organizational strategy and vision statements' (CSF1), 'Top management commitment and Leadership' (CSF2), 'Clear objectives, goals and responsibilities' (CSF5), 'Appropriate workforce skills and managers' expertise' (CSF6), 'Suitable selection and prioritization of projects and Technologies' (CSF8), 'Cyber security management' (CSF9).

4. Driving CSFs: In this cluster falls CSFs having strong driving power but weak dependence power. We found in this quadrant two CSFs, 'Employees awareness, training and involvement' (CSF3) and 'Allocation of resources and infrastructure' (CSF4).

6.5   RESULTS, DISCUSSION AND RESEARCH IMPLICATIONS

The transition to Industry 4.0 has become the major concern of researchers and industry professionals worldwide (Liao et al., 2017) due to the potential of digital technologies to increase process efficiency and operational excellence. In this context, it is suggested that LSS 4.0 has the potential to help manufacturing companies be more efficient. This current study examines the critical factors for successful implementation of LSS 4.0 through a mixed method approach. First, a systematic literature review was conducted to identify CSFs, which were then analyzed by a combination of a panel of experts from industry and academia and an ISM analysis to explore their interrelationships.

Ten relevant CSFs were identified from the literature and from discussions with experts in academia and industry. Two of these variables, CSF7 and CSF10, were determined to be dependent variables, two others as driving variables, which are CSF3 and CSF4, and six of them as linking variables. No CSF was identified as an autonomous variable. Building on the findings, the ISM model generated consists of six levels of hierarchy as shown in Figure 6.4. CSF10 was identified as the top level in the ISM model and CSF6 as the bottom level which is aligned with the results of (Samanta et al., 2023). It is observed from the model that 'Appropriate workforce skills and managers' expertise' held the first level in the model. Level 2 is represented by both 'Allocation of resources and infrastructure' and 'Employees awareness, training and involvement'. Further, level four is covered by four success factors, i.e. 'Clear objectives, goals and responsibilities', 'Top management commitment and Leadership', 'Suitable selection and prioritization of projects and Technologies' and 'Organizational, cultural and Technological readiness for LSS and I4.0'. The 'Appropriate skills and expertise of the workforce' CSF is at the ground level and the most vital CSF for the implementation of LSS 4.0.

To avoid failure and ensure success in LSS adoption, companies must have high leadership and commitment of top management. The advantages of adoption of LSS 4.0 in any organization, may encourage customers, suppliers, stakeholders and workforce to go ahead in the project. It is the most significant factor and plays a vital role in successful implementation of LSS 4.0. Unless this factor is strengthened, the focus on having other factors may not be effective. Next, there is a need to align and associate the LSS 4.0 with the organizational strategy, allocate appropriate and

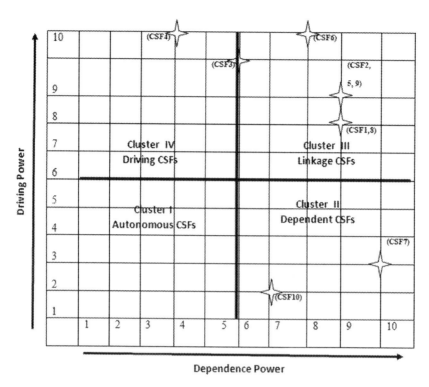

FIGURE 6.4   MICMAC analysis of CSFs.

skilled teams to the project by recruiting highly skilled and experienced employees, enhance workforce technical skills about emerging technologies by training and awareness campaigns and also provide the financial resource and infrastructure through equipment and consultancy. Organizations also need to set well-defined objectives and communicate them. Therefore, managers must assess and understand the company readiness level in terms of cultural, organizational and technological dimensions.

6.5.1   THEORETICAL AND PRACTICAL IMPLICATIONS

The results of this study are expected to contribute to both theoretical and practical implications by highlighting relevant CSFs to help researchers and practitioners design their journey to LSS 4.0.

In terms of theoretical implications, this study provides an overview and a prioritization of the critical success factors of LSS 4.0 as there has been little interest and work in the field of LSS 4.0 using multi-criteria decision making approaches. In addition, the study can serve as a guide for researchers in this field to explore this analysis and work on future perspectives proposed by the authors.

In terms of practical implications, our study identified the CSFs of LSS 4.0 and evaluated their interrelationships using an ISM approach. This research is useful for managers as it is based on expert opinions and experience, and establishes a prioritization of the CSFs affecting successful LSS 4.0 implementation. The use of ISM, a decision making tool-based approach will help managers to focus their strategic efforts to gain competitive advantage and to mitigate failure.

6.6 CONCLUSION, LIMITATIONS AND FUTURE RESEARCH PERSPECTIVES

This study presents the key factors for successful implementation of LSS 4.0. Initially an extensive literature review of relevant databases was performed based on a series of keywords to select articles dealing with LSS and Industry 4.0 success factors, which resulted in ten different factors being identified. In a second step, the list of selected CSFs was further discussed with a panel of experts.

Through this work, we contribute to an in-depth understanding of these success factors and their interdependencies by using an interpretive structural model to guide practitioners in their transition to LSS 4.0 in an effective manner. Ten critical success factors were identified and analyzed. Our results can be used to build a good digital transformation project, training programs, skills development, strategies and policies, etc. This research has two limitations. First, the results and the proposed model were developed based on the opinions of experts, which may present a bias. Second, as the research field is still in its early stages, the study presented ten CSFs for implementing LSS 4.0 that were identified from the literature and validated by the expert panel. We expect that as the field matures, several more CSFs may emerge.

To the best of our knowledge, within the LSS 4.0 research field, there are currently very few scientific studies examining CSFs with the ISM process or other decision making methods (Samanta et al., 2023). However, there are several research studies that focus on CSFs within the traditional context of Lean Six Sigma or Industry 4.0 separately (Belhadi et al., 2019; Cherrafi, Elfezazi, Chiarini, et al., 2017; Pozzi et al., 2021). This research is one of the first and early studies to analyze the CSFs of combining LSS and Industry 4.0 under the integrated approach. Our results are aligned with the results of (Samanta et al., 2023).

In this study, the authors used the ISM model. We suggest that further future research uses other techniques (e.g. ANP, AHP, DEMATEL, Fuzzy ISM) to unlock more results. Overall, the study provides useful insights into the critical success factors for implementing LSS 4.0 to avoid failure and promote its adoption. Therefore, it provides reliable evidence for manufacturing companies regarding the critical success factors that can facilitate the implementation of this approach. We suggest as future development on our study to execute an empirical test and validation of these results. The proposed model can be validated using hypothesis testing and structural equation models.

REFERENCES

Albliwi, S. A., Antony, J., & Lim, S. A. Halim. (2015). A systematic review of Lean Six Sigma for the manufacturing industry. *Business Process Management Journal, 21*(3), 665–691. https://doi.org/10.1108/BPMJ-03-2014-0019

Ali, S. M., Hossen, Md. A., Mahtab, Z., Kabir, G., Paul, S. K., & Adnan, Z. ul H. (2020). Barriers to lean six sigma implementation in the supply chain: An ISM model. *Computers & Industrial Engineering*, *149*. https://doi.org/10.1016/j.cie.2020.106843

Ali, Y., Younus, A., Khan, A. U., & Pervez, H. (2021). Impact of Lean, Six Sigma and environmental sustainability on the performance of SMEs. *International Journal of Productivity and Performance Management*, *70*(8), 2294-2318. https://doi.org/10.1108/IJPPM-11-2019-0528

Alqudah, S., Shrouf, H., Suifan, T., & Alhyari, S. (2020). A moderated mediation model of lean, agile, resilient, and green paradigms in the supply chain. *International Journal of Supply Chain Management*, *9*(4), 158–172.

Anass, C., Amine, B., Ibtissam, E. H., Bouhaddou, I., & Elfezazi, S. (2021). Industry 4.0 and Lean Six Sigma: Results from a Pilot Study. *Lecture Notes in Mechanical Engineering*, 613–619. https://doi.org/10.1007/978-3-030-62199-5_54

Antony, J., McDermott, O., Powell, D., & Sony, M. (2022). The evolution and future of lean Six Sigma 4.0. *The TQM Journal*, *35*(4), 1030–1047. https://doi.org/10.1108/TQM-04-2022-0135

Belhadi, A., Touriki, F. E., & Elfezazi, S. (2019). Evaluation of critical success factors (CSFs) to lean implementation in SMEs using AHP: A case study. *International Journal of Lean Six Sigma*, *10*(3), 803-829. https://doi.org/10.1108/IJLSS-12-2016-0078

Braun, V., & Clarke, V. (2006). Using thematic analysis in psychology. *Qualitative Research in Psychology*, *3*(2), 77–101. https://doi.org/10.1191/1478088706qp063oa

Briner, R. B., & Denyer, D. (2012). Systematic Review and Evidence Synthesis as a Practice and Scholarship Tool. In D. M. Rousseau (Ed.), *The Oxford Handbook of Evidence-Based Management*, 1ʳᵉ ed., 112–129. Oxford University Press. https://doi.org/10.1093/oxfordhb/9780199763986.013.0007

Buer, S.-V., Semini, M., Strandhagen, J. O., & Sgarbossa, F. (2021). The complementary effect of lean manufacturing and digitalisation on operational performance. *International Journal of Production Research*, *59*(7), 1976–1992. https://doi.org/10.1080/00207543.2020.1790684

Cherrafi, A., Elfezazi, S., Chiarini, A., Mokhlis, A., & Benhida, K. (2016). The integration of lean manufacturing, Six Sigma and sustainability: A literature review and future research directions for developing a specific model. *Journal of Cleaner Production*, *139*, 828–846. https://doi.org/10.1016/j.jclepro.2016.08.101

Cherrafi, A., Elfezazi, S., Chiarini, A., Mokhlis, A., & Benhida, K. (2017). Exploring Critical Success Factors for Implementing Green Lean Six Sigma. In L. Brennan & A. Vecchi (Eds.), *International Manufacturing Strategy in a Time of Great Flux*, 183–195. Springer International Publishing. https://doi.org/10.1007/978-3-319-25351-0_9

Cherrafi, A., Elfezazi, S., Garza-Reyes, J. A., Benhida, K., & Mokhlis, A. (2017). Barriers in Green Lean implementation: A combined systematic literature review and interpretive structural modelling approach. *Production Planning & Control*, *28*(10), 829–842. https://doi.org/10.1080/09537287.2017.1324184

Ciano, M. P., Strozzi, F., Minelli, E., Pozzi, R., & Rossi, T. (2019). The link between lean and human resource management or organizational behaviour: A bibliometric review. *Proceedings of the Summer School Francesco Turco*, 1 PartF, 321–328. www.scopus.com/inward/record.uri?eid=2-s2.0-85083665044&partnerID=40&md5=e50d33dbf79aecb541e9b8ac1822ac5

Denyer, D., & Tranfield, D. (2009). Producing a systematic review. *Undefined*. www.semanticscholar.org/paper/Producing-a-systematic-review.-Denyer-Tranfield/c451d2fdba4a42efcf074b718ca47c8a7f229edc

Drohomeretski, E., Gouvea da Costa, S., Pinheiro de Lima, E., & Garbuio, P. (2014). Lean, Six Sigma and Lean Six Sigma: An analysis based on operations strategy. *International Journal of Production Research*, *52*(3), 804–824. https://doi.org/10.1080/00207 543.2013.842015

Garza-Reyes, J. A. (2015). Lean and green – a systematic review of the state of the art literature. *Journal of Cleaner Production*, *102*(1), 18–29. https://doi.org/10.1016/j.jcle pro.2015.04.064

Ghobakhloo, M. (2020). Industry 4.0, digitization, and opportunities for sustainability. *Journal of Cleaner Production*, *252*. https://doi.org/10.1016/j.jclepro.2019.119869

Javaid, M., & Haleem, A. (2020). Critical components of Industry 5.0 towards a successful adoption in the field of manufacturing. *Journal of Industrial Integration and Management*, *5*(3), 327–348. https://doi.org/10.1142/S2424862220500141

Kumar, M. (2007). Critical success factors and hurdles to Six Sigma implementation: The case of a UK manufacturing SME. *International Journal of Six Sigma and Competitive Advantage*, *3*(4), 333. https://doi.org/10.1504/IJSSCA.2007.017176

Kumar, S., Luthra, S., Govindan, K., Kumar, N., & Haleem, A. (2016). Barriers in green lean six sigma product development process: An ISM approach. *Production Planning & Control*, 1–17. https://doi.org/10.1080/09537287.2016.1165307

Lameijer, B. A., Pereira, W., & Antony, J. (2021). The implementation of Lean Six Sigma for operational excellence in digital emerging technology companies. *Journal of Manufacturing Technology Management*, *32*(9), 260–284. https://doi.org/10.1108/JMTM-09-2020-0373

Laureani, A., & Antony, J. (2012). Critical success factors for the effective implementation of Lean Sigma: Results from an empirical study and agenda for future research. *International Journal of Lean Six Sigma*, *3*(4), 274–283. https://doi.org/10.1108/204014 61211284743

Liao, Y., Deschamps, F., Rocha Loures, E., & Ramos, L. (2017). Past, present and future of Industry 4.0—A systematic literature review and research agenda proposal. *International Journal of Production Research*, *55*(12), 3609–3629. https://doi.org/10.1080/00207 543.2017.1308576

Luthra, S., Yadav, G., Kumar, A., Anosike, A., Mangla, S. K., & Garg, D. (2020). Study of key enablers of industry 4.0 practices implementation using ISM-Fuzzy MICMAC approach. *Proceedings of the International Conference on Industrial Engineering and Operations Management*, 0(March), 241-251. www.scopus.com/inward/record.uri?eid=2-s2.0-8508 5482465&partnerID=40&md5=7f8f388ff76b4bdffe2f97a933f2678e

Moeuf, A., Pellerin, R., Lamouri, S., Tamayo-Giraldo, S., & Barbaray, R. (2018). The industrial management of SMEs in the era of Industry 4.0. *International Journal of Production Research*, *56*(3), 1118–1136. https://doi.org/10.1080/00207543.2017.1372647

Pozzi, R., Rossi, T., & Secchi, R. (2021). Industry 4.0 technologies: Critical success factors for implementation and improvements in manufacturing companies. *Production Planning & Control*, 1–21. https://doi.org/10.1080/09537287.2021.1891481

Ravi, V., & Shankar, R. (2005). Analysis of interactions among the barriers of reverse logistics. *Technological Forecasting and Social Change*, *72*(8), 1011–1029. https://doi.org/10.1016/j.techfore.2004.07.002

Samanta, M., Virmani, N., Singh, R. K., Haque, S. N., & Jamshed, M. (2023). Analysis of critical success factors for successful integration of lean six sigma and Industry 4.0 for organizational excellence. *The TQM Journal*, https://doi.org/10.1108/TQM-07-2022-0215

Shah, R., Chandrasekaran, A., & Linderman, K. (2008). In pursuit of implementation patterns: The context of Lean and Six Sigma. *International Journal of Production Research*, *46*(23), 6679–6699. https://doi.org/10.1080/00207540802230504

Skalli, D., Charkaoui, A., & Anass, C. (2022). *The Integration of Industry 4.0 in Operations Management: Toward Smart Lean Six Sigma*, 3–11. https://doi.org/10.1007/978-3-031-01942-5_1

Skalli, D., Charkaoui, A., & Cherrafi, A. (2022). Assessing interactions between Lean Six-Sigma, Circular Economy and industry 4.0: Toward an integrated perspective. *IFAC-PapersOnLine*, 55(10), 3112–3117. https://doi.org/10.1016/j.ifacol.2022.10.207

Skalli, D., Charkaoui, A., Cherrafi, A., Garza-Reyes, J. A., Antony, J., & Shokri, A. (2023). Industry 4.0 and Lean Six Sigma integration in manufacturing: A literature review, an integrated framework and proposed research perspectives. *Quality Management Journal*, 30(1), 16–40. https://doi.org/10.1080/10686967.2022.2144784

Snee, R. D. (2010). Lean Six Sigma – getting better all the time. *International Journal of Lean Six Sigma*, 1(1), 9–29. https://doi.org/10.1108/20401461011033130

Sony, M., Antony, J., Mc Dermott, O., & Garza-Reyes, J. A. (2021). An empirical examination of benefits, challenges, and critical success factors of industry 4.0 in manufacturing and service sector. *Technology in Society*, 67, 101754. https://doi.org/10.1016/j.techsoc.2021.101754

Sony, M., & Naik, S. (2020). Critical factors for the successful implementation of Industry 4.0: A review and future research direction. *Production Planning & Control*, 31(10), 799–815. https://doi.org/10.1080/09537287.2019.1691278

Tortorella, G. L., Giglio, R., & van Dun, D. H. (2019). Industry 4.0 adoption as a moderator of the impact of lean production practices on operational performance improvement. *International Journal of Operations & Production Management*, 39(6/7/8), 860–886. https://doi.org/10.1108/IJOPM-01-2019-0005

Yadav, G., & Desai, T. N. (2017). Analyzing Lean Six Sigma enablers: A hybrid ISM-fuzzy MICMAC approach. *The TQM Journal*, 29(3), 488-511. https://doi.org/10.1108/TQM-04-2016-0041

Yadav, N., Shankar, R., & Singh, S. P. (2020). Impact of Industry4.0/ICTs, Lean Six Sigma and quality management systems on organisational performance. *The TQM Journal*, 32(4), 815–835. https://doi.org/10.1108/TQM-10-2019-0251

7 Application of Industry 4.0, Digital Transformation, and Lean Six Sigma to Detect Cold Weld Defects

Jesús Vazquez-Hernandez,
Rosario Martínez-García,
Fernando González-Aleu,
Teresa Verduzco-Garza, and
Edgar M.A. Granda-Gutierrez

7.1 INTRODUCTION

Welding is a very versatile metallic confirmation procedure with the purpose of joining parts of metallic objects (Molera, 1992). Likewise, it is a highly complex process, which requires different components to be carried out, including temperature, technique, molds, machinery, time, and welders among others. Electrical Resistance Welding (ERW) is frequently used in steel pipe manufacturing organizations, and it refers to the method by which electrical current is passed through the weld. At the conclusion of the ERW process, a series of tests are carried out to determine the quality of the weld (e.g. metallographs). Sometimes, an irregularity that can appear is cold welding. Cold welding occurs during the process of joining two metals through ERW when there is not a correct fusion, and therefore a defect is generated. This irregularity occurs when there is roughness (crystal irregularity) in the tubes, in addition to having oxides present.

Metallographs are carried out to establish the state of the metal and its integrity, which is determined by the continuity or discontinuity of the metallic mass. When nonconformities of this type or similar occur, the investigation is carried out on the batches with suspected defects and they are withdrawn, becoming a loss for the company.

A 60 years' old international steel manufacturing pipe company (SMPC), with five manufacturing plants in Mexico and the U.S.A., satisfies customer demand from various industries, such as construction, automotive, and agriculture, among others. Currently, it has the most modern plant in Latin America for the manufacture of steel profiles, tubes, and components. Plant "A", placed in Mexico, has the biggest production capacity (more than 100,000 square meters and more than 1,300

DOI: 10.1201/9781003381600-7

workforce) with the newest technology. One year ago, Plant "A" received several claims and complaints for cold weld defects in rollover protective structures (ROPS), representing more than 32,000 USD in annual loss, plus the potential loss of these customers. Considering that Plant "A" is the most modern plant, leaders in the organization decided to begin its Industry 4.0 journey by implementing high-tech strategies such as digital information technologies.

Industry 4.0 is considered by many the fourth stage of the industrial revolution which involves the Internet of Things (IoT), Cyber-Physical Systems, digitalization, and other topics for processes, products, and services (Liao et al., 2017). Organizations implementing tools and methodologies related to Industry 4.0 reported several benefits, such as flexibility, efficiency, quality increase, and cost reduction (Masood and Sonntag, 2020). Since the introduction of Industry 4.0 in the practitioner and academic world, other derivatives of the "4.0" concept have been created, such as Quality 4.0 and Lean Six Sigma 4.0 (LSS 4.0). Quality 4.0 integrates quality management, improvement models, and technology (e.g. digital transformation and real-time data collection) to foster organizational competitiveness (Fonseca, Amaral, and Oliveira, 2021). The Quality 4.0 concept is currently included in the European Foundation for Quality Management 2020 model.

On the other hand, Gonzalez-Aleu and Garza-Reyes (2020) defined continuous improvement projects (CIPs) as multidisciplinary (typically) team-based processes used by organizations to improve their process performance metrics in a relatively short time (from days to months), such as Kaizen events, Six Sigma projects, Lean projects, Lean Six Sigma projects, and general quality improvement projects (Plan-Do-Check-Act). CIPs are a key factor in achieving organizational excellence or operational excellence. Therefore, LSS 4.0 is a CIP that consists of organizational process improvement using the DMAIC problem-solving methodology and fast information systems to achieve operational excellence (Arcidiacono and Pieroni, 2018; Sodhi, 2020). To the authors' knowledge, although LSS 4.0 appeared around 2018, there is a lack of publications describing CIPs applying DMAIC problem-solving methodology, Big Data (including real-time data collection), analytics, fast-making decision processes, and other technologies.

Hence, the purpose of this book chapter is to demonstrate how philosophies, methodologies, and tools from Industry 4.0, digital transformation, and LSS could be integrated to achieve organizational excellence. To reach this aim, the authors conducted CIP as an action research framework (some of the authors were directly involved in the CIP team) to improve cold weld defects detection in a steel pipe manufacturing company (SPMC) using DMAIC, CRISP-DM methodology (data mining methodology), analytics, software programming (Phyton), and metallographic picture analysis in real-time.

This book chapter is organized as follows. First, a literature review section about LSS 4.0 to identify key theoretical and practitioner contributions. Second, the action research methodology describes how DMAIC and CRISP-DM problem-solving methodologies were integrated and how data was collected. Third, results from each step of the action research methodology are documented. Lastly, in the discussion section, the authors summarize their findings, identify the main theoretical and practitioner contributions, address research limitations, and offer future research on LSS 4.0.

7.2   LEAN SIX SIGMA 4.0 LITERATURE REVIEW

According to Tissir et al. (2022), despite the growing interest in integrating I4.0 and LSS, only a few studies have attempted to synthesize and assess the related body of knowledge using standalone literature reviews. Sordan et al. (2021) performed a systematic literature review (SLR) which identified 13 contact points between I4.0 and LSS: (1) process mapping aided by cyber physical systems (CPS), big data analytics (BDA) and simulation; (2) performance measurement systems powered by CPS and BDA; (3) quality function deployment (QFD) powered by BDA; (4) capability analysis and statistical process control through CPS and BDA; (5) machine conditions monitoring aided by CPS, BDA and simulation; (6) optimization of handling and storage through CPS, simulation and AGV (Automated Guided Vehicle); (7) advanced robotics and CPS applied to manufacture cell design; (8) setup time reduction using modular structures and machine learning; (9) pull system aided by digital Kanban (e-kanban); (10) jidoka system and real time production control; (11) quality control powered by robotics and CMM; (12) mistake proofing solutions using sensing, AI, and augmented reality; and (13) augmented reality and e-learning applied to operational training.

Antony et al. (2022) defined four stages in the evolution of LSS. First, LSS 1.0, which consists of the integration of Lean and Six Sigma. Second, LSS 2.0 or LSS Green, included organizational environmental and sustainability impact. Third, LSS 3.0, or LSS Holistic, consists of improvement systems that create, sustain, and integrate improvement initiatives in any organization's culture and business environment. Fourth, LSS 4.0, integrates LSS practices and Industry 4.0 technologies (e.g., Internet of Things, big data, data analytics, digitalization, and others). Additional findings from Antony et al. (2022) include:

1. Benefits of LSS and Industry 4.0 integration: big data analytics in all DMAIC phases, real-time quality control, and event-based inspection.
2. Challenges of LSS and Industry 4.0 integration: lack of a framework to integrate LSS and Industry 4.0, digital operations could generate new waste (e.g., non-utilized talent and poor information management), frontline workers need to be highly skilled in decision-making, and some mapping tools need to be reinvented.

Practitioners and researchers in this field of LSS 4.0, suggest the following further lines of research: understand the impact of LSS 4.0 on corporate performance (Sordan et al., 2021; Antony et al., 2022), a framework that integrates LSS and Industry 4.0 (Chiarini and Kumar, 2021; Antony et al., 2022), sustainability of LSS 4.0 benefits (Antony et al., 2022), real case studies implementing LSS 4.0 (Sordan et al., 2021), and identified critical success factors for implementing LSS 4.0 (Yadav, Shankar, and Singh, 2021).

7.3   METHODOLOGY

To conduct this research, the authors integrated three research methodologies: action research (AR), CRISP-DM, and Lean Six Sigma (LSS). First, an AR is an interactive inquiry process that balances problem-solving actions implemented in a collaborative context with data-driven collaborative analysis or research to understand underlying causes, enabling future predictions about personal and organizational change (Gibertoni, Araujo and Menegon, 2016). AR has become consolidated effectively over the years, and it can be applied in different areas such as IT, education, social science, health, engineering, and others (Rowell et al., 2015). The researchers decided to use this research methodology instead of a case study because it was expected that at least one of the authors would be participating as a member of the CIP team. On the other hand, researchers in a case study will only be observers of the CIP (not a team member). Gonzalez-Aleu and Garza Reyes (2020) show AR methodology used in different CIPs, including the interaction between the CIP team, CIP facilitator, company facilitator, customer, and other stakeholders.

Second, released around the year 2000, the CRISP-DM (*Cross Industry Standard Process for Data Mining*) methodology for applying data mining projects consists of six iterative phases (Schoer, Kruse, and Gomez, 2021):

1. Business understanding. The business situation is assessed to identify resources required for the project, as well as their availability. Also, data security and regulatory criteria should be identified.
2. Data understanding. In this phase, data is collected from different sources and the data quality is assessed.
3. Data preparation. From the findings in the previous phase, bad data quality is handled by cleaning data.
4. Modeling. Depending on the business problem and data available, team members select the best modeling techniques.
5. Evaluation. Findings from the modeling phases are assessed against the business's initial objectives. According to this evaluation, the team members should define further actions.
6. Deployment. The project is documented as a final project report or a software component describing team findings and actions. This final report also needs to include actions and responsibilities related to the planning, deployment, monitoring, and maintenance of the improvement actions implemented.

Some authors recommend the application of CRISPM-DM in data mining projects that follow a structured plan-do-check-act (PDCA) cycle problem-solving methodology (Guruler and Istanbullu, 2014).

Third, LSS, which integrates Lean and Six Sigma philosophies to aid in increasing competitiveness, profitability, and process efficiencies, such as just-in-time, jidoka, and inference statistics analysis (Montgomery and Woodall, 2008; Sordan et al., 2021; Antony et al., 2022), using the DMAIC problem-solving approach:

1. Define. A team is assembled, a project chapter is created, one or more critical customer requirements are defined, and economic business impact is calculated.
2. Measure. In this phase, the measurement system is assessed for one or more critical quality characteristics.

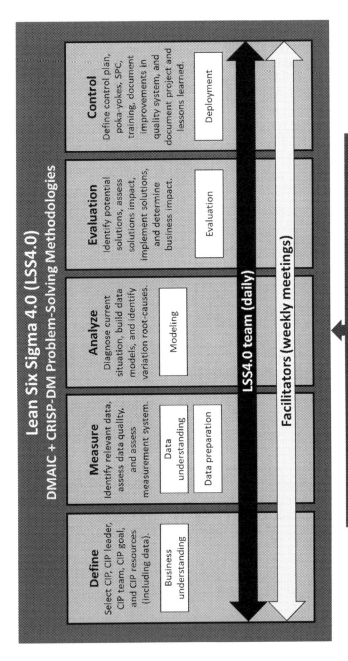

FIGURE 7.1 Lean Six Sigma 4.0 approach: DMAIC + CRISP-DM problem-solving methodologies.

3. Analyze. Data is analyzed to understand variation sources, process capabilities, and potential root causes.
4. Improvement. A list of potential solutions is created, evaluated, and implemented.
5. Control. A control plan and/or mistake-proof process is created to guarantee the sustainability of the actions implemented and the impact on the project objective.

With the advent of technological connectivity and access to massive data, the possibilities of augmenting the LSS DMAIC problem-solving approach with advanced technologies are enormous. Tay and Loh (2022) examined digital transformations (DT) of supply chains from a process improvement angle using the LSS DMAIC approach together with CRISP-DM.

Considering the nature of these three research methodologies and problem-solving approaches, the authors created the research framework shown in Figure 7.1.

7.4   RESULTS

The following section is structured using steps from the problem-solving methodology in Figure 7.1.

7.4.1   Define

The purpose of this phase was twofold: (i) define the problem and business current state; and (ii) define the LSS 4.0 project goals. These goals were achieved as shown in the following steps.

1. Business understanding. Plant "A" is the biggest manufacturing facility from a corporate group of five facilities with more than 60 years in the market. Plant "A" leaders were interested in beginning an Industry 4.0 transformation plan; therefore, the organization recently hired data analytics employees.
2. Problem definition. Cold welding occurs during the process of joining two metals through ERW, when there is not a correct fusion, and therefore a defect is generated. This irregularity occurs when there is roughness (crystal irregularity) in the tubes, in addition to having oxides present. At the time, Plant "A" did not have an anomaly detection system to ensure the quality of the products. To achieve quality products, destructive tests were performed, such as the cone test and the crush test. In addition, the irregularities, deformations, discontinuities, cracks, or openings inside the manufactured pipes and profiles were analyzed using metallography. Cold welding defects had been increasing since the second half of 2020 and were representing a problem, with claims from high-profile clients. This problem represented more than 32,000 USD annually, plus the potential loss of customers. Therefore, Plant "A" leaders decided to conduct an LSS 4.0 project that included topics such as increasing customer satisfaction, digital transformation, and analytics.

3. LSS 4.0 team. Plant "A" and an external consulting firm formed a team with five members: Senior Consultant as LSS 4.0 project leader (external), two Junior Consultants (external), one Data Analyst (Plant "A"), and one Process Engineer (Plant "A"). The three external team members were full-time focused on this LSS 4.0 project and the two Plant "A" team members were assigned for 30% of their time to the LSS 4.0 project. The LSS 4.0 team presented its progress every month with Plant "A" leaders.

4. LSS 4.0 facilitators. The LSS 4.0 team had support from facilitators and stakeholders. The role of the external facilitator was to guide the LSS 4.0 team in the application of DMAIC problem-solving methodology and the utilization of mathematical data modelling. The LSS 4.0 team reviewed its progress weekly with the facilitator. On the other hand, Plant "A" facilitator was available to the LSS 4.0 team to break departmental barriers and facilitate access to information and people in Plant "A".

5. LSS 4.0 timeline. Once the LSS 4.0 team was formed, roles, responsibilities, and due dates were defined. This project was expected to be completed in 16 weeks (around a four month period): Define phase (two weeks), Measure phase (two weeks), Analyze phase (six weeks), Improve phase (four weeks), and Control phase (two weeks).

6. LSS 4.0 project Scope. There are 25 mills used to manufacture steel pipes. According to the production per mill in kilograms, 81.36% of the production is focused in 40% of the mills. However, mill-ST is the machine with the most tonnage for manufacturing products under the rollover protection system (ROPS) classification. In 2020, mill-ST produced 82.8% of all the ROPS products in Plant "A". Quality is of vital importance in the products under this classification (ROPS) to guarantee the safety, health and working conditions of the operators in question (Ellis, 2020). It should be noted that a potential defect due to cold welding on this type of product would put the life of the operator at risk. Therefore, this LSS 4.0 project was focused on mills-ST.

7. LSS 4.0 project goal. At the time, Plant "A" did not have a test to identify cold welding defects in the process using metallography. Therefore, after several meetings with Plant "A" (representing the LSS 4.0 project customer), three goals were defined for this LSS project: (i) Design a classification model that allows the identification of cold welding defects with a performance level of 85% (gmean) in order to predict anomalies on the mill-ST; (ii) Determine the key factors influencing cold weld defects and quantify their correlation; and (iii) Implement and optimize the hyperparameters of the developed data analytics models.

Once the LSS 4.0 project goals were defined and approved, the LSS 4.0 team moved to the next DMAIC phase.

7.4.2 Measure

The purpose of this phase is to reduce the variation that the measurement system could introduce in the LSS 4.0 project, as well as to assess data quality. These goals were achieved as shown in the following steps.

1. Data understanding. Historical data was collected from two different databases to conduct this LSS project: the metallographic database that measures the quality of the products and the time series database (data visualization software) which shows variables throughout the production process. Metallographs are tests carried out by different operators and can be requested by various people within the organization. Additionally, specific software is used to develop the metallographs through microscope analysis for a better understanding of the images and possible welding defects. These studies are carried out by eight different laboratory technicians, of which only two are responsible for the results. Metallographic studies are handled as a type of destructive testing. This is because to carry out the study, it is necessary to cut a fragment of the revised piece, meaning it would no longer be useful for the client and it is discarded. On the other hand, dealing with the time series database, the totality of the data amounts to just over 195 million data records from July 2020 to August 2021. This information was initially in long format, indicating a record for each variable in the time stamp in which the sensors identify a change in the values.

2. Metallographic database preparation. To ensure the correct measurement of the metallographic images by the laboratory workers, an attribute R&R (Reproducibility and Repeatability) study was carried out, where the operators responsible for carrying out these metallographic tests were evaluated. Results from this study show that laboratory workers have a good level of effectiveness in the evaluation of metallographs and a low level of error rate. Dealing with the metallographic database includes identifying the data owner, understanding each variable from the information bank, and dealing with empty cells and data inconsistencies. A total of 2,631 metallographic samples were obtained from September 2020 to August 2021. From these samples, 34% were classified as non-conformity products.

3. Time series database preparation. The initial size of the database was obtained, and it was cleaned by implementing Phyton programming language codes. First, the unwanted columns were removed from the data frame. Consequently, it proceeded to deal with missing or erred data. Empty data records were found to exist within the time frame and value columns. These were removed to eliminate noise and ensure data quality. Subsequently, it was verified that there were no duplicate records. Lastly, the transformation of the data by scaling the time frame per second was implemented. This processing helps to interpret the information obtained from descriptive analytics in a better way. Additionally, it was decided to develop the descriptive statistics by variable, together with distribution graphs to analyze their behavior and finally choose those statistics that would be the input values for the classification model to be developed. From this analysis, there are two types of variables depending on their nature, discrete and continuous variables. The LSS 4.0 team decided that the criterion for the choice of statistics per variable would be the type of variable. Therefore, it was decided to obtain for each of the variables grouped by batch, the average, and the maximum when it was a discrete variable. On the other hand, for the continuous variables it

was decided to obtain, in addition to these two statistics, the standard deviation. Finally, the transformed database that shows the statistics per variable grouped by batch has a size of (17284, 83). That is, 17,284 different batches are shown, each one with 83 columns that refer to the variables expressed with their average, maximum, and sometimes the standard deviation. The database containing the metallography statuses per lot was directly linked to the database containing the statistics per lot described before.

4. Database junction. By joining the databases, it is sought that for each lot there is a status of accepted or rejected to obtain the input for the model and with said information, the model will learn to classify. This means that, when entering new information on the variables of a new batch, it can be evaluated and compared with the previous results already stored in the model to predict whether the metallography will be accepted or rejected. By doing this, only 1,581 metallographic sample records matched the batch records within the time series. Of these, 1,209 were accepted lots and 372 rejected.

Once both databases (metallographic and time series) were cleaned and integrated, the LSS 4.0 team began the analysis phase.

7.4.3 ANALYZE

This phase consists of the development of algorithms and mathematical programming to take advantage of and enhance the corresponding databases. In addition, within it, the modelling techniques relevant to the problem posed are selected and applied. It is important to mention that the Jupyter tool is used through Anaconda based on the Python language within work notebooks.

Of the proposed models, some were selected to develop the most pertinent models dealing with the nature of the problem. These classifier models are the following: Random Forest, Support Vector Machine, Logistic Regression, and K-Nearest Neighbours. The four models mentioned are identified as classification models with supervised learning. It is worth mentioning that within the development of each of these classifiers, different versions of them will be presented with the aim of improving their performance. Finally, from these developed models, the model that shows the greatest precision within its classification capacity in relation to the training results will be chosen.

Initially, the importance of the characteristics also known as feature importance was analyzed. This analysis was carried out to identify those variables that really contribute to the model. When mentioning this contribution, reference is made to the ability of the variable to explain the behavior of the target variable, in this case, the status of the metallography of the batch in question. Based on this analysis, the variables that did not contribute to explaining the behavior were discarded, as well as the variables above 0.90 of cumulative importance. This means that the variables that explain up to 90% of the behavior of the objective variable were selected. In this way, the number of variables that entered the model was reduced to 55. A sample of the importance and accumulated importance of the variables can be seen in Table 7.1.

TABLE 7.1
Variables, Importance Level, and Accumulated Importance Level (Sample)

Feature	Importance	Accumulated Importance
Variable 1 mean	0.0249	0.0249
Variable 2 mean	0.0239	0.0488
Variable 3 mean	0.0235	0.0723
Variable 4 mean	0.0226	0.0949
Variable 5 mean	0.0221	0.117
Variable 6 std	0.0213	0.1383
Variable 7 mean	0.0208	0.1591
Variable 8 mean	0.0203	0.1794
Variable 9 mean	0.0201	0.1995
Variable 10 mean	0.0197	0.2192
Variable 11 mean	0.0196	0.2388
Variable 12 mean	0.0194	0.2582
Variable 13 max	0.0187	0.2769
Variable 14 mean	0.0184	0.2953
Variable 15 max	0.0184	0.3137

In general, to describe the model development process, this can be explained first as the importation of the relevant libraries, as well as the reading of the database to be used. It is important to mention that for the K-Nearest Neighbors (k–NN) and Support Vector Machine models, the variables are initially scaled so that the scales used in each of the variables do not affect the model results.

Subsequently, the data will be divided into training and testing. Within the development of the model, it is vital to validate the data in question. For this, the method of splitting data into training and test sets, known as "train/test split", is used. By splitting the data into training and test sets, you may be holding back information from which the algorithm could learn. For this reason, a very large data set should not be assigned to the test set. However, the smaller this set is, the more inaccurate the model error estimate will be. When taking these issues into account, it is a question of balancing both trade-offs. It was decided to handle an 80:20 split for the training and test subsets respectively. This is because these parameters allow the model to have enough information to learn and at the same time have a good estimate of the error in the test subset.

After validating the data used within the training subset to train the developed model, the data set is tested for the evaluation of the model. This is done by applying the model to the training subset and then making predictions with the learning obtained from these data. These predictions are compared with the values of the test subset and the error is calculated based on this comparison. However, one of the most common mistakes in model development is overfitting. Overfitting the model would indicate the exact fitting of training data into statistical model. Generally, it is indicated that there is an overfit in the model if the difference between the R^2 of training and the R^2 of testing is greater than ten percentage points.

Once the above has been explored, the analysis is continued with the application of the method known as cross-validation. This method is used to be able to select a classification method empirically, that is, it is based on experience and observation of the data (Schaffer, 1993). After defining the implementation of the data division and cross-validation, we proceeded with the definition of the model through the import of the specific libraries of each of them.

Subsequently, the model is trained on the data with the training set. The precision of how well the model was trained is then evaluated by the training and testing accuracies. These accuracies calculate the R squared of each of the subsets. However, when there is a class imbalance as in this case (Accepted and Rejected), these R squared are not reliable, and they can only be used to assess whether there is an overfit in the model or not. Other important indicators that are used to evaluate the performance of the different classification models are the confusion matrix and the geometric mean (gmean). According to Shin (2020), a confusion matrix is a tool that shows the hits and misses in class classification. For this, the best performer searches for the totality of correct predictions.

After obtaining the results of the indicators explained above, the model is evaluated to see whether it obtained the desired performance level. If the outcome is negative, different techniques are implemented to improve its performance. One of these ways is by optimizing the hyperparameters used within Python's internal programming. It is worth mentioning that, of the selected models to be developed, it is only possible to optimize the hyperparameters of Random Forest, Support Vector Machine, and K-Nearest Neighbors. For Logistic Regression there is no such programming. To carry out this optimization, the optimizer is defined to obtain the best parameters subsequently. Python runs through all possible combinations of these parameters until the combination that results in the best classification performance is obtained. After this, the model is again defined and trained, now specifying the best parameters. Finally, the resulting model is evaluated with the aforementioned indicators.

Another technique that allows for improving the performance of the models is class balancing. For this, the SMOTEENN technique is generally used. This technique, according to Andhika (2021)

> combines the SMOTE ability to generate synthetic examples for the minority class and the ENN ability to remove some observations from both classes that are identified as having a different class between the observation class and its majority neighbour close K class.

This technique is only used in cases in which the original or optimized models do not result in the desired level of performance. This is because it performs over and under-sampling of data. In addition to this, another technique that allows for improvement in the performance of the models is the one known as SMOTE. This technique is used to perform oversampling and undersampling.

7.4.4 IMPROVE

Within this section, the evaluation of the models developed together with their variations is explained. These variations total 96 classifier models developed. To determine the optimal model that met the established objective, the team began by taking each of the four selected classifier models for the data corresponding to 1209 accepted batches and 372 rejected batches. Now, it is worth mentioning that to reach the desired result, it was necessary to carry out a series of iterations of each of these classifying models with different valid adjustments in order to find the best performance.

First, each of the four classifier models was scaled through the following scalers: Normalizer Scaler, Standard Scaler, and MinMax Scaler. Then, each of the scaled classifier models was divided into three techniques including hyperparameter optimization, SMOTEEN, and SMOTE. These combinations created thirty-six different scenarios.

When implementing the codes in Python, the confusion matrix was calculated internally together with its respective indicators such as the percentage of false negatives, false positives, and the gmean for each iteration. However, since the desired results were not obtained, a down-sampling suggested by experts within the company was carried out, where a total of 500 accepted lots and 372 rejected lots were contemplated to reduce the class imbalance. Again, the four classifying models, the three escalators, and the three techniques were divided, totaling thirty-six scenarios. Likewise, the confusion matrices with their respective indicators were calculated. However, none of the models resulted in the desired gmean.

Therefore, it was decided to obtain an egalitarian class balance, for which the data was divided into 372 accepted and 372 rejected batches. It should be noted that, of the total of 744 data, 595 are used for the train and 149 for the test. Once again, a down-sampling was carried out containing the four classifier models, the three scalers, and only two techniques: hyperparameter optimization (HO) and SMOTE. In this case, SMOTEEN is omitted because the technique uses oversampling and undersampling to balance classes, and this was done manually by equalizing the accepted and rejected batches. Therefore, there were only 24 possible scenarios. Then, the calculation of the confusion matrices and indicators was repeated.

Table 7.2 shows the sample of the results (showing those models scaled with Standard Scaler), where in the Random Forrest model row with the SMOTE technique,

TABLE 7.2
Models Evaluation

Model		True Negative	False Negative (Percentage)	True Positive	False Positive (Percentage)	gmean
kNN	HO	59	25 (17%)	50	15 (10%)	73%
	SMOTE	59	33 (22%)	42	15 (10%)	68%
Regression Logistic	HO	58	27 (18%)	48	16 (11%)	71%
	SMOTE	59	28 (19%)	47	15 (10%)	71%
Random Forest	HO	58	22 (15%)	53	16 (11%)	74%
	SMOTE	65	11 (7%)	64	9 (6%)	87%
SVM	HO	67	44 (30%)	31	7 (5%)	66%
	SMOTE	67	44 (30%)	31	7 (5%)	66%

a gmean of 87% is observed. The project objectives established that it was desired to develop a model with a performance of at least 85%, so the qualifying model meets the requirements. In the same way, the main criterion for the acceptance of the model in question is the confusion matrix, since a good performance is expected from the model. To qualify on this criterion, it is expected that there will be a low percentage of false negatives (7%) that are considered defects and false positives (9%) that are considered false alarms (see Table 7.2). Both percentages were expected to be in a range of less than 10% individually. The percentage was considered tolerable at the discretion of the company based on its experience in similar business problems.

Having mentioned the above, it can be established that the best model developed was accepted since the criteria established by the company were met. This model was considered the best model as it had the best performance in terms of the gmean metric. That said, the best classification model developed turned out to be the Random Forest classifier using the SMOTE technique and the Standard Scaler from the second down sampling applied.

Now, Figure 7.2 shows the confusion matrix that was generated to evaluate the performance of the classification model. The matrix compares the target-actual values with those predicted by the machine learning model (Akosa, 2017).

Likewise, the result of the accuracies of the best classifying model is presented. The training accuracy is 0.96, and the accuracy test has a result of 0.89. This means that the difference is less than ten percentage points, therefore, it is concluded that the model does not show overfitting.

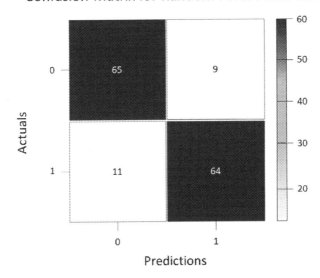

FIGURE 7.2   Confusion matrix for the Logistic Regression model developed with SMOTE and Standard Scaler.

TABLE 7.3
Model Test with Different Data Subsets

No.	Real Value	Predicted Value	Prediction Evaluation	Accepted Probability	Reject Probability
1	Accepted	Accepted	Correct	0.7241	0.2759
2	Rejected	Rejected	Correct	0.3467	0.6533
3	Rejected	Rejected	Correct	0.4239	0.5761
4	Accepted	Accepted	Correct	0.7586	0.2414
5	Rejected	Rejected	Correct	0.0301	0.9699

7.4.5 CONTROL

As part of the implementation, different data subsets were tested with the best developed model. This was done after applying the model on the training subset and making predictions with the learning obtained from these data on the test subset. A sample of these predictions can be seen in Table 7.3, where the real value, the predicted value, and the evaluation of the prediction are shown. All the observed samples resulted in a successful prediction due to the high gmean score of the model.

As part of the implementation or deployment in the company in question, various deliverables were presented to ensure that the proposed improvements were maintained within the company. The first of these was the optimized classification model. This refers to a file with its proper flow of steps for later use within the Python programming language. It is worth mentioning that in addition to this, the previous analysis and the other models developed were delivered in the same format to show the process of developing and choosing the best classifying model.

The second deliverable was a report on the results of the chosen model. The company received a report in a personalized format containing the results of the preliminary models, as well as the results of the models developed with the combined databases. Likewise, the results of the best selected model were explained in detail.

The third deliverable consisted of a report on recommendations to optimize the detection of anomalies. This report contained general recommendations to optimize the detection of anomalies or defects by cold welding within the pipeline. These recommendations included the use of the model, the welding process, and the improvement of the quality of evaluation of metallography, among others. Finally, the complete documentation of the present investigation was delivered. This documentation explained in detail the investigation process from the contextualization of the problem to the results of the developed model.

7.5 CONCLUSIONS

It can be established that a classification model was successfully designed that allowed the prediction of anomalies, and identification of defects due to cold weld defects. The best model developed obtained a performance calculated with a gmean of 87%. Therefore, the objective was successfully met in terms of performance.

The main reason why it was decided to develop this model was a high increase in claims from one of the most high-profile clients of the steel manufacturing company. It is worth mentioning that within the industry there is no tool as such that can detect cold weld defects within the process. Therefore, the present investigation was considered a practical contribution, evidencing its functionality within the operation. On the other hand, from a theoretical perspective, this research contributes to the body of knowledge related to integrating quality methods and Industry 4.0 technologies at the operational level (Fonseca, Amaral, and Oliveira, 2021; Antony et al., 2022). First, the LSS 4.0 team agreed that project success was related to having project goal clarity, following step by step the LSS 4.0 methodology (DMAIC + CRSIP-DM), and LSS 4.0 team members including the project target area employees. These critical success factors are similar to those mentioned by Gonzalez-Aleu and Van Aken (2016). However, an additional critical success factor included an expert in data analytics. Second, a barrier or challenge during this LSS 4.0 project was data cleaning and database integration; the two activities highlighted demanded time (about four weeks). These aspects could be related to data availability and trustworthiness (Gonzalez-Aleu and Van Aken, 2022).

As mentioned before, one of the most relevant limitations of this investigation was the accessibility and purification of information in terms of the available databases. To mitigate these limitations, exhaustive data processing and engineering were carried out, together with data transformation and data wrangling. Further research should follow four lines. First, the results of this research are within the plan for the development of more robust models that integrate new process variables within the steel manufacturing company. This is with the aim of developing a complete anomaly detection system to ensure product quality. Second, conduct a similar investigation, using the same model developed in different industries that seek specific applications with the aim of reducing the defect rate and ensuring the quality of their products. Third, additional case studies and action research publications on LSS 4.0 are needed to understand the new skills, knowledge, and abilities that LSS 4.0 project participants (facilitators, stakeholders, and team members) should develop. Lastly, additional research should be conducted to understand the success factors, barriers, and challenges in LSS 4.0 projects.

REFERENCES

Akosa, J. (2017). Predictive accuracy: A misleading performance. Oklahoma State University, pp. 12.

Andhika, R. A. (2021). Imbalanced classification in python: Smote-enn method. https://towardsdatascience.com/imbalanced-classification-in-python-smoteenn-method-db5db06b8d50.

Antony, J., McDermott, O., Powell, D., and Sony, M. (2022). "The evolution and future of Lean Six Sigma 4.0," *The TQM Journal*, pp. 1–18.

Arcidiacono, G. and Pieroni, A. (2018). "The revolution Lean Six Sigma 4.0," International Journal of Advanced Science Engineering Information Technology, 8(1), pp. 141–149.

Chiarini, A. and Kumar, M. (2021). "Lean Six Sigma and Industry 4.0 integration for Operational Excellence: evidence from Italian manufacturing companies," *Production Planning and Control*, 32(13), pp. 1084–1101.

Ellis, S. (2020). What does rops mean on a tractor? Farm Animals. https://farmandanimals.com/what-does-rops-mean-on-a-tractor/.

Fonseca, L., Amaral, A., and Oliveira, J. (2021). "Quality 4.0: The EFQM 2020 Model and Industry 4.0 Relationships and Implications," *Sustainability*, 13(6), pp. 3107–3127.

Gibertoni, D., de Araújo F.T., and Menegon, N. L. (2016). "The contribution of action research in the construction of scientific knowledge in Brazilian Production Engineering," *Production*, 26(2), pp. 373–384.

Gonzalez-Aleu, F., Garza-Gutierrez, D., Granda-Gutierrez, E.M.A. and Vazquez-Hernandez, J. (2022). "Increasing Forklift Time Utilization in a Food Equipment Manufacturing Plant with a Kaizen Event," In Advances in Manufacturing III: Volume 3-Quality Engineering: Research and Technology Innovations, Industry 4.0, 1(1), pp. 182–193.

Gonzalez-Aleu, F. and Garza-Reyes, J.A. (2020). *Leading continuous improvement projects: Lessons from successful, less successful, and unsuccessful continuous improvement case studies*. London: CRC Press

Gonzalez-Aleu, F. and Van Aken, E. (2016). "Systematic literature review of critical success factors for continuous improvement projects," *International Journal of Lean Six Sigma*,7(3), pp. 214–232.

Guruler, H. and Instanbullu, A. (2014). "Modeling student performance in higher education using data mining. In: Peña-Ayala, A. (eds) *Educational Data Mining. Studies in Computational Intelligence*, Springer.

Liao, T., Deschamps, F., Lourdes, E.D.F.R., and Ramos, L.F.P. (2017). "Past, present and future of Industry 4.0: A systematic literature review and research agenda proposal," *International Journal of Production Research*, 55(12), pp. 3609–3629.

Masood, T. and Sonntag, P. (2020). "Industry 4.0: Adoption challenges and benefits for SMEs," *Computer in Industry*, 121, pp. 1–12.

Molera, P. (1992). *Soldadura industrial: clases y aplicaciones.* Barcelona: Marcombo.

Montgomery, D.C. and Woodall, W.H. (2008). "An Overview of Six sigma," *International Statistical Review*, 76(3), pp.329–346.

Rowell, L.L., Polush, E.Y., Riel, M., Bruewer, A. (2015). "Action researchers' perspectives about the distinguishing characteristics of action research: a Delphi and learning circles mixed-methods study," *Educational Action Research*, 23 (2), pp. 243–270.

Schaffer, C. (1993). "Selecting a classification method by cross-validation". https://link.springer.com/article/10.10072FBF00993106.

Schroer, C., Kruse, F., and Gomez, J.M. (2021). "A Systematic Literature Review on Applying CRISP-DM Process Model," *Procedia Computer Science,* 181, pp.526–534.

Shin, T. (2020). "Comprensión de la matriz de confusión y cómo implementarla en python." www.datasource.ai/es/data-science-articles/comprension-de-lamatriz-de-confusion-y-como-implementarla-en-python.

Sodhi, H.S. (2020). "When Industry 4.0 meets Lean Six Sigma: a review," *Industrial Engineering Journal*, 13(1), pp. 1–12.

Sordan, J.E., Oprime, P.C., Pimenta, M.L., da Silva, S.L., and González, M.O.A. (2021). "Contact points between Lean Six Sigma and Industry 4.0: A systematic review and conceptual framework," *International Journal of Quality & Reliability Management*, 39(9), pp. 2155–2183.

Tay, H.L. and Loh, H.S. (2022). "Digital transformations and supply chain management: a Lean Six Sigma perspective," Journal of Asia Business Studies, Vol. 16 No. 2, pp. 340–353. https://doi.org/10.1108/JABS-10-2020-0415

Tissir, S., Cherrafi, A., Chiarini, A., Elfezazi, S., and Bag, S. (2022). "Lean Six Sigma and Industry 4.0 combination: Scoping review and perspectives," *Total Quality Management & Business Excellence*, pp. 1–30.

Yadav, N., Shankar, R., and Singh, S.P. (2021). "Critical success factor for lean six sigma in quality 4.0," *International Journal of Quality and Service Sciences,* 131(1), pp. 123–156.

8 Application of Lean Six Sigma 4.0 in Seed Potato Value Chain Performance Improvement

Gurraj Singh

8.1 INTRODUCTION

The agricultural sector of the developing world is facing huge competition as the rate of globalization increases. Under such circumstances these countries face direct competition from their more developed counterparts in the same target markets. The globalization of product features, quality and supply, as well as the distribution network, pose an enormous challenge to both the logistics as well as the supply chain in the agricultural sector of those nations, mostly from South East Asia, Sub Saharan Africa and Latin America [1]. As a consequence, it is becomingly increasingly necessary for the producers to improve and regulate their own trade and distribution. along with getting a better control on the production process as well. Additionally, these producers also need to enact measures to meet up to the rigorous safety and quality regulations of such markets [2]. Importantly, small-scale producers encounter most of these issues in difficult circumstances, due to possession of diminutive capital. Lack of institutional and infrastructural control along with the unavailability of basic resources adds to the plight of the producers of such countries [3]. On one hand, the most notable characteristic of these global value chains is the collapse of international barriers, as a consequence of falling tariffs alongside export subsidies. However, on the other hand, the growing dominance of the western powers in the global market, highlighting the asymmetric power relationships, is discouraging the producers from these developing countries from entering such markets [4]. Apart from numerous other similar barriers like market access, orientation and institutional voids, another notable factor is the unavailability of resources and infrastructure for the small growers. The most important contributors to this particular factor are the possession of knowledge and capabilities. It is the lack of these factors, alongside proper skills and technological access, that lead to the aggravation of the issue [5].

Focusing specifically on the scenario of agriculture in India, as per the annual report by the Department of Agriculture, Cooperation and Farmers Welfare, over half of the total population of India is directly or indirectly dependent on agriculture for their livelihood [6]. Moreover, as per a report published by the Food and Agriculture

DOI: 10.1201/9781003381600-8

Organization (FAO) of the United Nations, over 80% of the farmers in India have been categorized as "small scale" farmers [7]. India possesses a total of 329 million hectares of geographical area, out of which over 50% is directly under agriculture. Being a sub-continent within itself, the country is fragmented into numerous agro-climatic zones (15 zones), typically on the basis of the availability of water resources, rainfall patterns, and soil quality as well as the average temperatures throughout the year [6][8]. However, the scenario has become gloomy over the past few decades and the ever-rising number of suicides has tainted the image of the Indian agricultural sector over the decades. These are known to account for over 11% of all the suicides in India. As per the National Crime Records Bureau of India, there were over 13,000 cases of reported farmer suicides in the country in the year 2014 [9]. There are numerous factors which may be held accountable for these including indebtedness, failure of monsoons, government policies, personal/family issues etc. Nevertheless, most of these factors can be directly or indirectly linked to the lack of economic viability in agricultural practices, especially when implemented on a small scale [10]. These disadvantaged farmers, due to their small land holdings and lack of technical skills, face a huge challenge in selling their produce in most of the markets, which as discussed in the previous section are dominated by their western counterparts. This often leads to diminishing profits or even losses at times, which over a few years evolve into a hefty debt leading to the unfortunate eventuality of these suicides [11][12]. This vicious circle cannot simply be dealt with at the farmer's end alone, since government policies are the major steering force in deciding the fate of the farmer. However, the application of various industrial engineering techniques like Lean Six Sigma (LSS), focusing on the production and marketing processes, can act as a welcome contributor in a whole-hearted effort to assist and support the troubled farming sector of India and on similar lines, the rest of the developing world.

With the COVID-19 pandemic clearly exposing the highly fragile nature of the global infrastructure, techniques like LSS are swiftly gaining popularity as front runners for application to the agricultural sector. They can either be applied in optimizing the yields for better revenue or for the effective minimization of unused assets like labor or machinery. These improvements can effectively eliminate wasteful activities by the simultaneous application of LSS principles, thus improving the quality of the produce as well. A recent study by Laux and Sabharwal (2018) applied LSS principles in crop storage and post-harvest handling applications based in the Latin American country of Columbia. The researchers have been able to expose the lack of work done in the related domains of food security and farm management in a systematic manner [13]. Another study by Ariffien et al. (2021) focused on the post-harvest handling of vegetables in Indonesia [14]. The application of Value Stream Mapping (VSM) helped in visualizing the respective lengths of the value-added and the non-value-added processes. Eventually, a 171.46 min reduction in packaging time was achieved, thus ensuring better quality, timely delivery and assured profits for the farmers. Another detailed research study by Joshi et al. (2017) explicitly analysed the agricultural scenario in India. It concluded that industrial engineering techniques, like LSS, can be highly effective in the optimization of resources through harmonious utilization [15].

Another vital element at small scale farms that needs a special mention is the association between the farmer and the laborer (since small scale farms are more labor intensive). Improved coordination can improve the quality of produce while simultaneously reducing the lead times. In a related study, but applied to dairy farming, Melin and Bath (2019) used VSM to formulate an action plan at a small dairy farm in Sweden. It was observed that the use of VSM not only created a collaborative culture among the human resources but also led to significant scientific improvements in the milk quality as well as the animal health [16]. A similar application to various agricultural practices such as growing, harvesting etc. could play a vital role in improving the overall efficiency of the process.

Potato (Solanum Tuberosum) is considered as one of the most essential staple food crops throughout the world. It is second only to maize when talking in terms of the number of countries growing them (79% of the world's countries grow potato). In terms of tonnage, it ranks fourth after wheat, rice and maize, thus making it an important vegetable. Comprising of 80% water, 18% starch and 2% proteins, potato is a very significant, cheap source of calories to meet the energy requirements of the people living in poor countries.

In context of the Indian sub-continent, potato cultivation was strongly encouraged by the British during the pre-independence era. Presently, India stands as one of the leading producers of potato. The potato seed is produced within the country and negligible imports are made. The northern plains of India i.e. the Jalandhar district and some parts of the Kapurthala district of Punjab have been acknowledged as the most suitable seed production areas in India by the Central Potato Research Institute (CPRI), Shimla, India [17]. Potato farming in India is very diverse and complicated due to the difference in climatic, socio-economic, agronomic and cultural conditions. The division of the Indian subcontinent is made on the basis of the different agro-economic, latitudinal, longitudinal, altitudinal, climatic, soil, and irrigation conditions etc. The northern plains of India grow two crops of potato i.e. autumn and spring [18]. The central and southern parts of India with some hilly areas grow a summer crop as well. On average the summer crop is hardly 20% of the autumn crop. Also, the autumn crop, grown according to the seed plot technique, yields the seed potato in the Jalandhar and Kapurthala districts of Punjab [19] [20]. The seed produced in these areas is used by a major portion of the remaining Indian states for sowing both the spring as well as the autumn crop.

Based upon the discussions in the previous section, the goal of the current research is to apply the lean production concepts in order to reduce waste activities in the supply chain of seed potato from the Jalandhar District of Punjab, India. The novelty of the current study lies in its effort to use the LSS approach in making the seed potato supply chain more efficient and robust to eventually benefit both the grower and the end customer. The study will primarily focus on the management of logistical issues in the supply chain through an analytical framework and operational coordination. The novelty of the current work, additionally and more importantly, is to achieve added-value as well as competitive advantage, thus empowering the producers with a better placement when competing with the bigger players in the potato seed industry, including the likes of Mahindra HZPC and ITC Chambal. It is therefore aimed at preventing the current supply chain from moving INTO the hands of these corporate giants, which is essential for the success of the small-scale growers.

8.2   RESEARCH METHODOLOGY

The current study has made use of VSM as a LSS tool with the main working model based on the seed potato supply chain in the Jalandhar District of the Punjab state of India, which is a major potato seed production hub in the country. A case study has been conducted by collecting quantitative as well as qualitative data, and additionally exploring other complex issues. The collection of information, which is the backbone of this study, was done through multiple channels including field visits, in-depth interviewing, and analyzing related documents as well as the available literature. The interviewing was done through 20 informants to obtain information about the day-to-day activities in the production, harvest, post-harvest etc. processes. Each of the informants was a potato grower with land holdings no larger than five hectares. This was done to ensure the capture of data related to small farmers, who are the prime focus of this study.

It is a well-established fact that the VSM tool helps in reducing waste, costs, and lead times, and also streamlines the work process [21]. The process uses a series of steps to successfully design and implement the model. The very first step comprises of the drawing of the current state map with the assistance of the collected data. It is this current mapping that is indispensable for formulating and drawing the future state map after a careful consideration of the principles of LSS. The drawing of the future mapping necessarily takes into account various useful principles, like targeting a continuous flow or the balancing of the process by minimizing the waiting time. Similarly, the improvement in quality can be simply achieved by successfully visualizing a defect and identifying the several factors responsible for it using an Ishikawa (cause and effect) diagram. Eventually, the combination of all such improvements is compiled in the most practical and financially viable manner to present the new value stream process that will replace the old one.

8.3   CURRENT STATE MAPPING

The process of analyzing the current map of the seed potato supply chain reveals that there is a series of dissimilar sequences of events throughout the supply chain. The current study focuses primarily on the activities involving production, harvest and delivery of the seed potatoes. To analyze the status quo, it is very important to be aware of the various stakeholders/parties involved in the supply chain. A detailed map of the flow of information and finances in the current supply chain has been depicted in Figure 8.1. The farmer acts as the producer or the grower of the commodity under study, i.e. the seed potato. The farmer himself is in charge of all the activities ranging from production to marketing his produce to either big farmers or local traders who then supply the commodity to the interstate supplier who is also a private trader. The farmer isn't directly able to sell his produce to the interstate

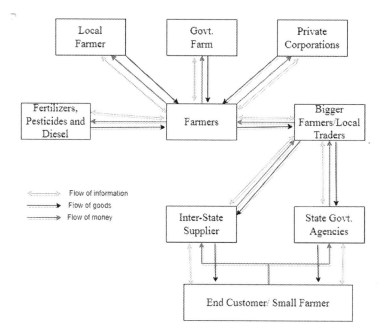

FIGURE 8.1 Seed potato supply chain characteristics.

supplier due to his inability to amass the threshold quantities of the commodity viable for interstate transportation over very long distances. He is therefore forced to sell his produce locally despite knowing that the end customer is also a small-scale farmer but in another distant region of the country. It is therefore these interstate private traders who supply the seed to the end customers. The farmer is able to procure seed from three alternative sources, namely other farmers, government farms or from the private corporations. The government farms are capable of supplying a very limited quantity of new seed, which reaches only 5% of the total farmers. This is mainly due to the lack of infrastructure developed by the various government agencies for this purpose. On the other hand, the big corporations, like the ones mentioned earlier, are able to supply good quality seed, but at strikingly high prices. Therefore, the best alternative available to the farmer is the other local farmers or traders who supply medium or in some cases, sub-standard seed at lower prices. Similarly, the fertilizer shop provides the farmers with the necessary chemical protection and nutrient supplements for their crop, while the diesel is procured from the local gas stations run by either semi-government or private corporations. It would also be important to highlight another supplier i.e. the state government agencies, which also procure seed from Jalandhar, Punjab, through official tendering procedures. However, the small farmers are again not able to reap the benefits of better prices due to their inability to supply a certain threshold quantity. Figure 8.2 depicts a SIPOC diagram that includes the suppliers, inputs, processes, outputs and customers, thus helping a better understanding of the process. It has been clearly observed that there are a series of value-adding processes, and also that there is no effort from the farmer to maintain the quality and assure it

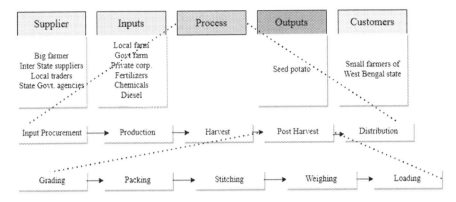

FIGURE 8.2   SIPOC mapping of the production process of seed potato.

consistently. The main reason for this is a very short supply period, which is 2–3 weeks before the weather gets unbearably hot across all of northern India in the first week of May.

The average customer demand is 25 tons per week per farmer. However, it can increase with the increase in supply as the orders are accepted on a first come first serve basis. Careful examination of the seed potato value stream reveals that there is a lot of time wastage in activity completion, thus accounting for the high overall costs. The current value chain uses no mechanization at all and the consequence of this is visible in the low levels of quality control that exists among average farm households.

As far as the information flow is concerned, the customers contact the farmer telephonically. The rates are usually decided as per the market trends in the present, as well as the future. The fixing of rates is an entirely informal process. The local farmers who are currently the focus of the study, usually get poor rates due to the absence of proper grading, resulting in varying sizes, as well as the inability to sort out the defected/cut/bruised potatoes, which further reduces the level of quality. The current seed production value stream as depicted in the VSM in Figure 8.3. has the gross time for VA (value-added) processes as 49.1 hours while the gross NVA (non-value-added) process time is 130.9 hours. Overall, the VA activities account for 27.2% of the total cycle time, while the remainder 72.7% falls under the NVA processes. This is mainly due to traditional harvesting and post-harvesting management practices, alongside the faulty hand grading methods being used by the local farmers. As per the VSM diagram, the 120 hours delay before the grading is a wasteful activity which is mainly due to the thought process of the farmer to firstly harvest the entire crop and keep on accumulating the inventory (produce). and grading only after the harvesting is over. Meanwhile the harvested potato is piled up in huge quantities in open fields and covered with paddy hay. This also reduces the quality of the crop, especially its color, due to daily perspiration and interaction with the soil sticking to its skin. This factor has a huge impact on the final price at which the potatoes are sold. Once the harvest is over, the crop is then loaded and taken to the warehouse where it is graded using manual labor which is usually untrained. This further reduces the quality, thus

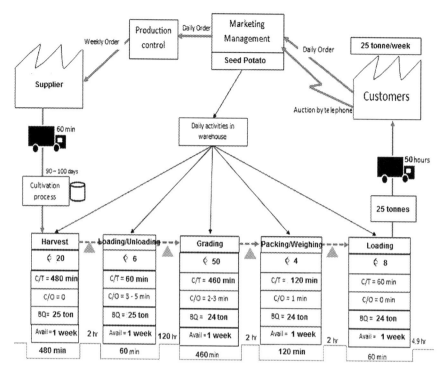

FIGURE 8.3   Current state map of seed potato value stream.

contributing to sub-standard quality produce, which can be highlighted as one of the major reasons for the farmers getting low rates as compared to the bigger farms/corporations which implement mechanization. Thus, it can be concluded that the current value chain is loaded with time wasting activities which can be avoided, and the overall result can be improved.

8.4   FUTURE STATE MAPPING

On the basis of the primary data available in the current state map, the future state map has been formulated using the analyses of several processes including customer requirement and flow, as well as production planning. It is a well-known fact that the demand of the customer in this value chain can vary with the variation in supply by the farmer. As a result, the future map converts the existing pull system into a push system, in which the farmer is able to supply two loads of 25 tons per week. The plan seeks to introduce mechanization into the current model especially during the harvesting and grading processes. Inexpensive locally manufactured implements and machines should be introduced into the process as per the plan. Additionally, once the harvest is loaded into the tractor-trailer, it is unloaded directly into the size grader which has a twin benefit of using one-fifth the amount of labor and improving the final

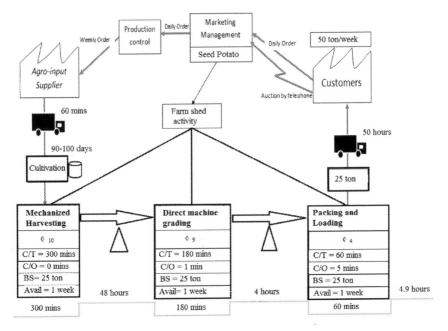

FIGURE 8.4   Future state map of seed potato value stream.

product quality by many fold. Moreover, the inclusion of cut/bruised potato is also avoided. The detailed representation of the future state VSM is shown in Figure 8.4.

The work cell 1 represents the harvesting, grading and loading combinations. Due to the usage of a mechanized potato digger, the labor costs as well as the cycle times are bound to come down substantially to 300 mins as compared to the 480 mins in the current case. The work cell 2 is the most important one with a combination of grading and quality improvement. An important advantage is that the product need not be taken to the warehouse, instead the grader can be carried to the field and the process can be performed without any intermediate transportation. Moreover, the cost of the grader can be recovered in only two seasons, in light of the drastic reduction in the labor costs involved in manual grading of the produce. The cycle time also decreases to 180 mins from 460 mins per hectare of produce. Finally, work cell 3 involves the packing, stitching and loading of the packed produce into the transport vehicle. The total VA cycle time for the process is expected to be 22.5 hours while the total NVA cycle time drastically reduces to 56.9 hours from 130.9 hours in the current state scenario. Also, the percentage of VA activities may increase from 27.2% to 35.2%. Hence, the improvement in the future state production system involves the usage of low-cost mechanization in the existing process, with a main focus on harvesting and grading methods. As a result, the frequency of weekly loading is expected to double, which along with improved quality may increase the profit margins of the farmer. The introduction of the push system favors the farmer since the risk of overheating and damaging of the produce due to an advancing summer is averted.

8.5   CONCLUSION

The seed potato production and marketing system in Jalandhar, Punjab, India can be improved further by several important steps. The intervention of state governments in supplying subsidized potato diggers and size graders can play a leading role in achieving these goals. In order to maximize the operational process, the dumping of the produce after harvesting is to be avoided and replaced by direct grading with an automatic or a semi-automatic grader. Furthermore, there are still certain issues that may be addressed to improve the overall functionality of the potato seed production and distribution supply chain. It would need an all-round involvement from the government agencies, firstly in supplying superior quality seed to the farmer to improve his yields, and secondly in assisting the farmer in the inter-state marketing and supply of seed potatoes by acting as a mediator between the farmer and the government agencies of the other states. Additionally, distribution administration along with improved levels of mechanization will play a pivotal role in achieving an effective, efficient as well as a robust supply chain.

REFERENCES

[1] J H Trienekens (2011) "Agricultural Value Chains in Developing Countries A Framework for Analysis", *International Food and Agribusiness Management Review* 14(2): 51–82.

[2] A De Janvry, E Sadoulet (2005) "Achieving success in rural development: toward implementation of an integral approach", *Agricultural economics* 32(s1): 75–89.

[3] B Daviron, P Gibbon (2002) "Global Commodity Chains and African Export Agriculture", *Journal of agrarian change* 2(2):137–161.

[4] G Gereffi, J Humphrey, T Sturgeon (2005) "The governance of global value chains", *Review of International Political Economy* 12(1): 78–104.

[5] E Giuliani, C Pietrobelli, R Rabellotti (2005) "Upgrading in global value chains: Lessons from Latin Americal clusters", *World development* 33(4): 549–573.

[6] Annual Report (2020–21) Department of Agriculture, Cooperation and Farmers Welfare, Ministry of Agriculture and Farmers Welfare, Government of India.

[7] G Rapsomanikis (2015) "The economic lives of smallholder farmers", *Food and Agriculture Organization of the United Nations*, Rome, 2015.

[8] All India Report on Agriculture Census (2015–16) Department of Agriculture, Cooperation & Farmers Welfare, Ministry of Agriculture & Farmers Welfare, Government of India, 2020.

[9] G L Parvathamma (2016) "Farmers Suicide and Response of the Government in India - An Analysis", *IOSR Journal of Economics and Finance* 7(3):1–6.

[10] K R S Sravanth, N Sundaram (2019) "Agricultural Crisis and Farmers Suicides in India", *International Journal of Innovative Technology and Exploring Engineering* 8(11): 1576–1580.

[11] B B Mohanty (2013) "Farmer Suicides in India", *Economic and political weekly* 48(21): 45–54.

[12] "Every Thirty Minutes: Farmer Suicides, Human Rights, and the Agrarian Crisis in India," Center for Human Rights and Global Justice, New York: NYU School of Law (2011).

[13] C Laux, R Sabharwal (2018) "The Application of Lean Six-Sigma in Food Security and Food Safety: A LSS strategy for Small Holder Farmers", *Seventh International Conference on Lean Six Sigma, 7th & 8th May 2018*

[14] A Ariffien, I Adriant, J A Nasutian (2021) "Lean Six Sigma Analyst in Packing House Lembang Agriculture Incubation Center (LAIC)", *Journal of Physics: Conference Series* 1764 (2021) 012043 doi:10.1088/1742-6596/1764/1/012043

[15] A Joshi, W Bankar, V Deshpande (2017) "Application of Lean Six Sigma to Indian Farming Sector", *International Conference on Quality Up-gradation in Engineering, Science & Technology(ICQUEST-2017)*.

[16] M Melin, H Barth (2020) "Value stream mapping for sustainable change at a Swedish dairy farm", *Int. J. Environment and Waste Management* 25(1): 130–140.

[17] ICAR Sponsored Winter School on Advancement in Potato Production Technology and its Future Prospects (November 19 to December 9 (2019), ICAR-Central Potato Research Institute.

[18] V Bhardwaj, S Rawat, J Tiwari et al. (2022) "Characterizing the Potato Growing Regions in India Using Meteorological Parameters", *Life* 12, 1619. https:// doi.org/ 10.3390/life12101619

[19] R K Singh, T Buckseth, V Singh, R Kumar, S K Chakrabarti (2022) "Seed Potato Production in India", *Current Horticulture: Improvement, Production, Plant Health Management and Value-addition.* www.indianjournals.com/ijor.aspx?target= ijor:chr&volume=9&issue=2&article=br

[20] R K Singh, T Buckseth, J K Tiwari, et al. (2019) "Seed potato (Solanum tuberosum) production systems in India: A chronological outlook", *Indian Journal of Agricultural Sciences* 89 (4): 578–87.

[21] T Rohac, M Januska (2015) "Value stream mapping demonstration on real case study", *Procedia Engineering* 100:520–529.

9 Unleashing the Potential of Industry 4.0 in India

Opportunities, Challenges, and the Road Ahead

Bikram Jit Singh and Harsimran Singh Sodhi

9.1 INTRODUCTION

Industry 4.0, also known as the fourth industrial revolution, refers to the integration of advanced digital technologies into manufacturing processes to drive innovation, improve efficiency, and enhance product quality (Dos Santos et al., 2021). This transformational approach to manufacturing is driven by the convergence of technologies such as artificial intelligence, the Internet of Things, robotics, and cloud computing (Kumar et al., 2020). The potential of Industry 4.0 to drive economic growth and create jobs has made it a key focus for governments and industry leaders around the world (Bongomin et al., 2020). In India, the government has recognized the importance of Industry 4.0 for the country's economic growth and has taken several initiatives to support its adoption.

India has a strong manufacturing sector that contributes significantly to the country's economy, accounting for nearly 15% of its GDP. The adoption of Industry 4.0 technologies has the potential to transform India's manufacturing sector, making it more competitive in the global market and driving growth and job creation (Sodhi, 2021). In recent years, several Indian companies have successfully implemented Industry 4.0 technologies, demonstrating the potential of these technologies to drive innovation and enhance productivity (Ustundag and Cevikcan, 2018). However, there are also challenges that need to be addressed to drive the widespread adoption of Industry 4.0 in India, such as the need for reliable infrastructure, a skilled workforce, and a favorable regulatory environment.

The world is witnessing the fourth industrial revolution, also known as Industry 4.0, which is the integration of digital technology and advanced manufacturing techniques (Jabbar et al., 2022). India, being one of the largest economies in the world, is slowly adopting Industry 4.0, and there are several opportunities and challenges that come with it. This book chapter will explore the challenges and opportunities for deploying Industry 4.0 technologies in India, as well as the driving factors, case studies, and future prospects for the technology. The chapter will also provide insights into the readiness of India to deploy Industry 4.0 technologies and explore the dire need for some robust operational excellence approaches like Lean Six Sigma 4.0 (LSS 4.0) to create the prerequisites to support Industry 4.0 sustainably.

9.2   WHY INDUSTRY 4.0?

Industry 4.0 is needed to address the challenges and opportunities presented by the current global economic landscape (Mavropoulos and Nilsen, 2020). Here are some of the key reasons why Industry 4.0 is necessary. Industry 4.0 technologies such as IoT, AI, and advanced robotics can help to automate and optimize industrial processes, resulting in increased efficiency and productivity. This is particularly important as global demand for products and services continues to grow, and companies need to produce more with the same or fewer resources (Lambrechts et al., 2021). By increasing efficiency, Industry 4.0 technologies can help companies to reduce costs (Kaliraj and Devi, 2022).This is important as companies face pressure to maintain profitability in an increasingly competitive global marketplace. Industry 4.0 technologies can enable companies to create more customized products and services. With the rise of e-commerce and online marketplaces, consumers increasingly expect personalized products and experiences. Industry 4.0 technologies can help companies to meet these expectations. Industry 4.0 technologies can provide companies with real-time data on their operations, allowing them to make better-informed decisions (André, 2019). This can help companies to identify inefficiencies and make changes to their operations quickly. By implementing Industry 4.0 technologies, companies can stay competitive in the global marketplace. This is particularly important as emerging economies such as India and China continue to grow and compete with established economies (Nath et al., 2020).

9.3   GLOBAL SUCCESS STORIES

Here are some global case studies of successful Industry 4.0 implementations:

Siemens AG: German manufacturing company Siemens AG has been a leader in the adoption of Industry 4.0 technologies. The company has implemented data analytics and the Internet of Things (IoT) to improve the efficiency of its manufacturing operations and reduce costs. Siemens has also implemented advanced robotics and automation to improve the quality of its products (Seth et al., 2022a).

BMW: German automobile manufacturer BMW has implemented Industry 4.0 technologies such as robotics and automation to improve the efficiency of its manufacturing operations (Kadir et al., 2018). The company has also implemented data analytics and the IoT to improve the quality of its products and reduce waste.

Haier: Chinese appliance manufacturer Haier has implemented Industry 4.0 technologies such as data analytics and the IoT to improve its production processes and reduce costs. The company has also implemented a modular manufacturing system that allows it to quickly adapt to changing customer demand (Singh and Rathi, 2019).

General Electric: US-based industrial conglomerate General Electric (GE) has implemented Industry 4.0 technologies such as data analytics and the IoT to improve the efficiency of its manufacturing operations. GE has also implemented predictive maintenance to reduce downtime and improve the reliability of its equipment (Seth et al., 2022b).

Airbus: European aircraft manufacturer Airbus has implemented Industry 4.0 technologies such as robotics and automation to improve the efficiency of its manufacturing operations. The company has also implemented data analytics to improve the quality of its products and reduce waste (Singh et al., 2020).

These case studies demonstrate that Industry 4.0 technologies can be successfully implemented in companies around the world, from German manufacturing giants like Siemens and BMW to Chinese appliance manufacturers like Haier. By leveraging advanced technologies such as robotics, automation, data analytics, and the IoT, these companies have been able to improve their efficiency, reduce costs, and remain competitive in the global market.

9.4 LIMITATIONS OF THE EXISTING INDIAN MANUFACTURING SCENARIO

Before the adoption of Industry 4.0 technologies, the Indian manufacturing sector faced several bottlenecks that hindered its growth and competitiveness. Here are some of the key bottlenecks. India's manufacturing sector suffered from an inefficient supply chain that led to delays in production and higher costs. The lack of infrastructure such as good roads, ports, and logistics facilities made it difficult to transport raw materials and finished goods (Fatorachian and Kazemi, 2021). India's manufacturing sector suffered from low productivity due to a lack of skilled labor, poor working conditions, and outdated equipment. This resulted in lower quality products, increased waste and higher costs (Safar et al., 2020). Quality control was often lacking in India's manufacturing sector, leading to products that did not meet customer expectations. Ineffective operational excellence further resulted in a loss of customer confidence and reduced demand for Indian products (Luthra and Mangla, 2018).

India's manufacturing sector lacked access to advanced technologies such as automation, robotics, and artificial intelligence. This made it difficult for manufacturers to optimize their operations, resulting in reduced efficiency and productivity (Sarkar et al., 2023). Complex regulatory environment: India's regulatory environment was often seen as complex and bureaucratic, making it difficult for manufacturers to navigate. This resulted in delays in obtaining permits and licenses, and increased compliance costs (Sharma and Singh, 2022). Many small and medium-sized enterprises in India's manufacturing sector faced a lack of funding, which made it difficult to invest in new equipment and technologies. This limited their ability to compete with larger companies and restricted their growth potential.

Overcoming these bottlenecks is necessary to create a more competitive manufacturing sector in India (Dutta et al., 2020). The adoption of Industry 4.0 technologies has the potential to address many of these bottlenecks, and to transform India's manufacturing sector into a more efficient, productive, and competitive industry.

However, the adoption of Industry 4.0 technologies requires significant investment in technology, infrastructure, and human capital, as well as changes to the regulatory environment.

9.5 EFFICACY OF INDUSTRY 4.0 IN INDIA

The adoption of Industry 4.0 technologies in India has the potential to transform the country's manufacturing sector (Lödding et al., 2017). The Indian government has recognized the potential of Industry 4.0 and has launched several initiatives to promote its adoption, such as the "Make in India" initiative. According to a report by Deloitte, the adoption of Industry 4.0 in India could increase the country's GDP by $957 billion by 2035 (Deloitte, 2018). India has a large labor force, but its productivity lags behind that of other countries. The adoption of Industry 4.0 technologies such as AI and robotics can help to automate and optimize manufacturing processes, resulting in increased efficiency and productivity (Iyer, 2018).

The adoption of Industry 4.0 technologies can also help to reduce costs in the Indian manufacturing sector. India is a price-sensitive market, and manufacturers face pressure to keep costs low (Sun et al., 2022). The adoption of Industry 4.0 technologies can help manufacturers to reduce costs while maintaining quality. The use of Industry 4.0 technologies can also help Indian manufacturers to create more customized products and services. With the rise of e-commerce and online marketplaces, consumers increasingly expect personalized products and experiences. The use of emerging technologies such as AI can help manufacturers create products that are tailored to the specific needs of individual consumers. One of the challenges that the Indian manufacturing sector faces is the lack of skilled labor (Shanker et al., 2019). The adoption of Industry 4.0 technologies requires a skilled workforce. Indian manufacturers must invest in training and development programs to ensure that their employees have the necessary skills to operate and maintain these technologies (Parhi et al., 2022).

The implementation of these technologies can improve the efficiency and productivity of the sector, reduce costs, and enable manufacturers to create more customized products and services (Saxena et al., 2022). The Indian government's initiatives to promote the adoption of Industry 4.0, combined with the country's large labor force in various campaigns like Make in India, Made in India etc.

9.6 CHALLENGES TO DEPLOYING INDUSTRY 4.0 IN INDIA

Despite the potential benefits of Industry 4.0 technologies for India's manufacturing sector, there are several challenges to deploying these technologies (Government of India, 2019). Here are some of the key challenges:

Limited digital infrastructure: India's digital infrastructure is still developing, with many rural areas lacking basic internet connectivity (Chauhan et al., 2021). This limits the ability of companies to deploy Industry 4.0 technologies that rely on a robust digital infrastructure.

Skills gap: There is a significant skills gap in India's workforce, particularly in the areas of advanced technology such as robotics, automation, and artificial intelligence. Without a skilled workforce, companies will struggle to deploy and effectively utilize Industry 4.0 technologies (Primi, A., & Toselli, M. (2020).

High costs: The cost of implementing Industry 4.0 technologies can be prohibitively expensive for many small and medium-sized enterprises in India. These companies may lack the necessary resources to invest in the technology and infrastructure needed to deploy Industry 4.0 solutions (Siqueira and Davis, 2021).

Resistance to change: Many traditional manufacturers in India may be resistant to adopting Industry 4.0 technologies due to a fear of change and a lack of understanding about the benefits of these technologies (Sarkar et al., 2022). This may result in slow adoption of these technologies and a lag in the industry's overall competitiveness.

Data security concerns: As Industry 4.0 technologies rely heavily on data and connectivity, data security concerns are a major challenge for companies in India. There is a lack of clear regulations and standards around data security, and many companies may not have the resources to invest in robust cybersecurity measures (Asghar et al., 2020).

Regulatory challenges: The Indian regulatory environment can be complex and bureaucratic, which may make it difficult for companies to obtain the necessary permits and licenses to deploy Industry 4.0 technologies. This could result in delays and increased costs for companies looking to adopt these technologies (Ada et al., 2021).

Lack of standardization: The lack of standardization across different Industry 4.0 technologies and platforms can make it difficult for companies to integrate and utilize these technologies effectively (Chow, 2019).

Overcoming these challenges will require a concerted effort from industry leaders, policymakers, and educators. Addressing the skills gap, providing access to funding, and creating a supportive regulatory environment will be crucial for the successful deployment of Industry 4.0 technologies in India. Integration of Lean Six Sigma (LSS) with Industry 4.0 is a relatively new concept that has been gaining momentum due to its potential to improve operational efficiency, product quality, and customer satisfaction. However, there are some challenges that need to be overcome to achieve successful integration. In this section, some of the challenges to integrate LSS with Industry 4.0 are discussed.

One of the main challenges of integrating LSS with Industry 4.0 is the lack of skills and expertise required to implement both systems. LSS requires a specialized skill set in statistical analysis, problem-solving, and project management, while Industry 4.0 requires expertise in digital technologies such as the Internet of Things (IoT), Artificial Intelligence (AI), and big data analytics (BDA). The challenge is to find individuals who possess both skill sets. Industry 4.0 relies heavily on data integration and connectivity to enable real-time data analysis and decision-making. However, integrating data from different sources can be challenging due to differences in data formats and standards. This can lead to data quality issues and affect the accuracy of LSS analysis. Industry 4.0 involves connecting various devices and machines to the internet, which increases the risk of cyber-attacks. Implementing LSS in such an environment can be challenging as data security is critical for maintaining the integrity of the analysis.

Integrating LSS with Industry 4.0 requires a cultural shift in the organization, and this can be challenging. The traditional LSS approach focuses on improving efficiency and reducing waste, while Industry 4.0 emphasizes innovation and digital transformation. Therefore, organizations need to balance the two approaches to achieve successful integration. The integration of LSS and Industry 4.0 can be expensive, especially for small and medium-sized enterprises (SMEs). The cost of implementing both systems can be a significant barrier for SMEs, and this can limit their ability to compete with larger organizations.

In a nutshell, the integration of LSS and Industry 4.0 has the potential to improve operational efficiency and product quality. However, it is not without its challenges. Organizations need to address these challenges and find ways to overcome them to achieve successful integration.

9.7   ENABLERS FOR INDUSTRY 4.0 IN INDIA

There are several driving factors and enablers that are promoting the implementation of Industry 4.0 in India. These include:

Government initiatives: The Indian government has launched several initiatives to support the adoption of Industry 4.0 technologies, including the "Make in India" program, which aims to boost domestic manufacturing, and the "Digital India" program, which aims to promote the development of digital infrastructure (Goswami and Daultani, 2022). These initiatives provide funding, tax incentives, and other support to companies looking to implement Industry 4.0 technologies.

Availability of skilled workforce: India has a large pool of young and tech-savvy talent, which provides a significant advantage for the implementation of Industry 4.0 technologies. There are also several educational institutions in India that offer courses and training programs in areas such as robotics, automation, and data analytics (Kumar et al., 2023).

Access to capital: India has a thriving startup ecosystem, with a growing number of venture capitalists and angel investors willing to fund startups and small businesses looking to adopt Industry 4.0 technologies (Muller and Kazantsev, 2021).

Improved digital infrastructure: India has made significant progress in developing its digital infrastructure, with the adoption of technologies such as 4G and the development of the National Optical Fiber Network. This has improved connectivity and made it easier for companies to deploy Industry 4.0 technologies that rely on a strong digital infrastructure.

Increasing demand for advanced manufacturing: As India seeks to become a global manufacturing hub, there is a growing demand for advanced manufacturing technologies that can improve efficiency and reduce costs. Industry 4.0 technologies are seen as essential to meeting this demand.

Increasing competition: India's manufacturing sector is facing increasing competition from other countries such as China, Vietnam, and Thailand. To remain competitive, Indian manufacturers must adopt advanced technologies such as Industry 4.0 to improve their efficiency, reduce costs, and improve product quality.

Support from industry associations: Industry associations such as the Confederation of Indian Industry (CII) and the Federation of Indian Chambers of Commerce and Industry (FICCI) are promoting the adoption of Industry 4.0 technologies and providing support to companies looking to implement these technologies (Rahimian et al., 2021).

Overall, these driving factors and enablers are promoting the adoption of Industry 4.0 technologies in India and are likely to accelerate the pace of implementation in the coming years. A cultural shift is essential to ensure the integration of Lean Six Sigma with Industry 4.0. It is important to create a culture of continuous improvement that values innovation and experimentation, promotes collaboration, and encourages risk-taking. This culture will help organizations to embrace change and integrate new technologies into their existing processes. The success of Lean Six Sigma and Industry 4.0 integration depends on the knowledge and skills of the workforce. Companies need to train their employees in both Lean Six Sigma and Industry 4.0 technologies to ensure that they can successfully implement and operate these tools. This training should include not only technical skills but also soft skills such as communication, problem-solving, and critical thinking.

Effective data management is crucial to the success of Lean Six Sigma and Industry 4.0 integration. Organizations need to ensure that their data is accurate, accessible, and secure. They should also develop systems and processes for managing and analyzing data to identify areas for improvement and inform decision-making. Integrating Lean Six Sigma with Industry 4.0 requires the use of advanced technologies such as Artificial Intelligence (AI), Big Data, and the Internet of Things (IoT). These technologies should be integrated into existing processes to improve efficiency and effectiveness. The successful integration of Lean Six Sigma and Industry 4.0 requires support from leadership and management. They should create a vision for integration, provide resources, and encourage experimentation and innovation. Collaboration and partnership between organizations and across industries is essential to integrating Lean Six Sigma with Industry 4.0. This collaboration can help organizations share knowledge, resources, and best practices to drive innovation and improvement.

Industry 4.0 includes a range of innovative technologies like Internet of Things (IoT), cloud computing, cybersecurity, simulation, autonomous robotics system etc. that allow the value chain to minimize production lead times and increase the quality of the product and organizational efficiency. Each Industry 4.0 technology is linked with a crucial role that is particular to its functioning in the manufacturing sector. Figure 9.1 depicts the six vital technologies (out of nine emerging ones) of Industry 4.0.

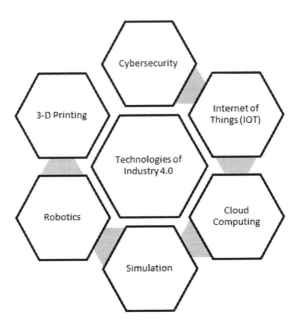

FIGURE 9.1   Important technologies associated with Industry 4.0.

9.8   INDIAN CASE STUDIES

There are several Indian companies that have successfully implemented Industry 4.0 technologies to improve their operations and drive business growth. Here are some case studies:

Mahindra & Mahindra: Mahindra & Mahindra (2021), a leading Indian automobile manufacturer, has implemented Industry 4.0 technologies such as robotics and automation to improve efficiency and quality in its manufacturing operations. The company has also implemented data analytics and the Internet of Things (IoT) to improve visibility of its supply chain and reduce costs. (Mahindra & Mahindra, 2021).

Godrej & Boyce: Godrej & Boyce, a leading Indian manufacturing company, has implemented Industry 4.0 technologies such as robotics and automation to improve productivity and reduce costs. The company has also implemented data analytics to improve its quality control processes and reduce waste (Suresh et al., 2018).

Hero MotoCorp: Hero MotoCorp, one of India's largest two-wheeler manufacturers, has implemented Industry 4.0 technologies such as automation, IoT, and data analytics to improve its production processes and reduce costs. The company has also implemented predictive maintenance to reduce downtime and improve the reliability of its equipment (https://cio.economictimes.indiatimes.com/news/internet-of-things/how-hero-motocorp-is-prepping-itself-for-industry-4-0/49477833).

Tata Steel: Tata Steel, a leading Indian steel manufacturer, has implemented Industry 4.0 technologies such as robotics, automation, and data analytics to improve the efficiency of its manufacturing operations (Kamble et al., 2018). The company has also implemented predictive maintenance to reduce downtime and improve the reliability of its equipment.

Bosch India: Bosch, a global technology company with operations in India, has implemented Industry 4.0 technologies such as IoT and data analytics to improve the efficiency of its manufacturing operations (Jena and Patel, 2022). The company has also implemented predictive maintenance to reduce downtime and improve the reliability of its equipment.

These case studies demonstrate that Industry 4.0 technologies can be successfully implemented in Indian companies across a range of industries, from automobiles to steel production. By leveraging advanced technologies such as robotics, automation, data analytics, and the IoT, these companies have been able to improve their efficiency, reduce costs, and remain competitive in the global market.

9.9   DISCUSSION OF INDIAN READINESS TO DEPLOY INDUSTRY 4.0

India is at various stages of readiness to deploy Industry 4.0 technologies across its industries. While some sectors and companies have made significant progress in implementing these technologies, others are still in the early stages of adoption (Hajoary and Akhilesh, 2021). One of the key enablers of Industry 4.0 is the availability of reliable and high-speed internet connectivity. While India has made significant progress in this area in recent years, there are still challenges in terms of the availability of high-quality infrastructure in remote or rural areas (Çınar et al., 2021). The successful deployment of Industry 4.0 technologies requires a highly skilled workforce that is capable of understanding and operating these technologies. While India has a large pool of talented engineers and technology professionals, there is a need for more training and upskilling programs to develop the specialized skills required for Industry 4.0 (Sony and Aithal, 2020). The Indian government has taken several initiatives to support the adoption of Industry 4.0 technologies, such as the National Policy for Advanced Manufacturing and the "Digital India" program. However, there are still regulatory challenges that need to be addressed, such as the lack of a clear data protection framework and outdated labor laws (Rockenschaub, 2016). Industry collaboration is essential to drive the adoption of Industry 4.0 technologies. While there are several industry associations and initiatives in India that are working towards this goal, there is still a need for greater collaboration across sectors and between academia and industry (Tripathi and Gupta, 2021).

Despite these challenges, there are several success stories of Industry 4.0 implementation in India, as discussed in previous sections (Aithal and Sony, 2020). The Indian government and industry leaders recognize the importance of Industry 4.0 for the country's economic growth and are taking steps to overcome the bottlenecks and drive adoption of these technologies. With the right policies, infrastructure, and collaboration, India has the potential to become a global leader in Industry 4.0.

9.10 VITAL RECOMMENDATIONS

Based on the analysis presented in this chapter, here are some recommendations for the successful implementation of Industry 4.0 in India. To enable the widespread adoption of Industry 4.0 technologies, India needs to develop reliable and robust infrastructure, including high-speed internet connectivity, advanced data centers, and digital platforms for collaboration. Industry 4.0 requires a highly skilled workforce that is proficient in advanced technologies such as Artificial Intelligence, robotics, and machine learning. India needs to invest in training programs to develop this skillset and create a pipeline of talent for the future. Beside these, following suggestions are made for developing excellence in India through Industry 4.0:

Operational Excellence: Operational excellence is essential for businesses undergoing digital transformations in the context of Industry 4.0. Industry 4.0 technologies, such as the Internet of Things (IoT), Artificial Intelligence (AI), and robotics, generate massive amounts of data that can be used to optimize production and operations. Operational Excellence methodologies, such as Lean Manufacturing, Six Sigma, and Total Quality Management, can help organizations use this data to identify inefficiencies, streamline processes, and reduce waste, resulting in increased efficiency and productivity. Lean Six Sigma is a set of business management strategies that combines Lean and Six Sigma methodologies to improve the efficiency and quality of manufacturing processes. With the advent of Industry 4.0, Lean Six Sigma 4.0 has emerged as a methodology that combines Lean Six Sigma with Industry 4.0 technologies to optimize manufacturing processes and improve overall business performance. Lean Six Sigma is a methodology that aims to reduce waste and variability in manufacturing processes while maintaining or improving quality standards. By applying Lean Six Sigma principles, manufacturers can reduce the time, cost, and defects associated with their processes. In the context of Industry 4.0, Lean Six Sigma 4.0 combines the principles of Lean Six Sigma with the latest digital technologies, such as Big Data Analytics (BDA), the Internet of Things (IoT), and Artificial Intelligence (AI). One of the key benefits of Lean Six Sigma 4.0 is its ability to optimize processes through real-time data analysis. In a study by Yoon et al. (2019), the authors demonstrated how Lean Six Sigma 4.0 can be used to optimize the production process of a steel mill. By collecting real-time data from the production line and applying Lean Six Sigma principles, the researchers were able to reduce the variability in the process and improve the quality of the final product. Another advantage of Lean Six Sigma 4.0 is its ability to detect and prevent defects before they occur. In a case study by Rao and Roy (2018), the authors demonstrated how Lean Six Sigma 4.0 can be used to improve the quality of automotive parts. By combining Six Sigma with IoT technologies, the researchers were able to detect defects in real-time and prevent them from occurring in future production runs. Furthermore, Lean Six Sigma 4.0 can help reduce the time and cost associated with the production process. In a study by Kumar and Tiwari (2018), the authors showed how Lean Six Sigma

4.0 can be used to optimize the supply chain process. By applying Lean Six Sigma principles to the supply chain and incorporating Industry 4.0 technologies, the researchers were able to reduce lead times, minimize inventory costs, and improve overall efficiency.

Flexibility: Industry 4.0 technologies enable businesses to be more flexible and responsive to changing market demands. Operational Excellence can help organizations achieve this flexibility by reducing lead times, increasing production flexibility, and improving supply chain agility (Jahani et al., 2021).

Cost savings: Digital transformations can be costly, and it is essential to ensure that investments are being used efficiently. Operational Excellence methodologies can help organizations optimize their investments by identifying opportunities to reduce costs and improve efficiencies.

Quality: Industry 4.0 technologies can enable businesses to improve product quality and reduce defects. Operational Excellence methodologies can help organizations achieve this by implementing robust quality control measures and continuous improvement programs.

Safety: Industry 4.0 technologies can introduce new safety risks, such as cyber threats and the potential for human-robot interactions. Operational Excellence can help organizations manage these risks by implementing effective safety protocols and training programs.

In summary, Operational Excellence is critical for businesses undergoing digital transformations in the context of Industry 4.0. By optimizing processes, improving efficiency and productivity, and reducing costs and risks, businesses can take full advantage of the benefits offered by Industry 4.0 technologies.

Collaboration and partnerships: Industry collaboration and public-private partnerships can play a critical role in driving the adoption of Industry 4.0 technologies. The government should encourage collaboration between industry players, academia, and research institutions to promote innovation and knowledge-sharing.

Regulatory framework: The government should create a favorable regulatory environment that promotes the adoption of Industry 4.0 technologies. This includes policies that incentivize the adoption of advanced technologies, protect intellectual property, and ensure data privacy and security.

Funding and investment: To support the adoption of Industry 4.0 technologies, the government should create a conducive environment for funding and investment. This can include tax incentives, funding for research and development, and support for startups and small businesses.

Standards and interoperability: Standards and interoperability are crucial for the seamless integration of different Industry 4.0 technologies. India should work towards the development of common standards and protocols that enable interoperability between different technologies and platforms.

By focusing on these recommendations, India can successfully implement Industry 4.0 technologies and drive innovation, growth, and job creation in the years to come.

9.11 CONCLUSIONS

The adoption of Industry 4.0 technologies in India can bring several opportunities, including increased efficiency, cost reduction, and a skilled workforce. However, the adoption of Industry 4.0 technologies comes with several challenges, including infrastructure, cost, security, and a skilled workforce. Companies and the government need to work together to overcome these challenges and adopt Industry 4.0 technologies to remain competitive in the global market (Neville and Doyle-Kent, 2022). In conclusion, the scope of Industry 4.0 in India is vast, with the potential to transform the country's manufacturing sector and contribute to its economic growth. Industry 4.0 technologies can drive innovation, improve efficiency, reduce costs, and enable Indian companies to remain competitive in the global market.

India has several strengths that can enable the successful deployment of Industry 4.0 technologies, such as its large pool of talented engineers and technology professionals, growing entrepreneurial ecosystem, and supportive government policies. Several Indian companies have already implemented Industry 4.0 technologies successfully and are reaping the benefits.

However, there are several challenges that need to be addressed to drive the widespread adoption of Industry 4.0 in India. These challenges include the need for reliable infrastructure, a skilled workforce, and a favorable regulatory environment. Industry collaboration and public-private partnerships can play a crucial role in overcoming these challenges and driving the adoption of Industry 4.0 technologies in India.

Overall, the successful deployment of Industry 4.0 technologies can have a significant impact on the Indian economy, driving growth and job creation. With the right policies, infrastructure, collaboration, and operational excellence techniques like LSS 4.0, India can become a global leader and drive innovation and growth in the years to come. In conclusion, Lean Six Sigma 4.0 can be a powerful tool for improving the efficiency, quality, and profitability of manufacturing processes in the context of Industry 4.0. By combining the principles of Lean Six Sigma with the latest digital technologies, manufacturers can optimize their processes in real-time, prevent defects before they occur, and reduce the time and cost associated with production. As Industry 4.0 continues to evolve, Lean Six Sigma 4.0 is likely to become an essential methodology for manufacturers looking to stay competitive in the global marketplace.

REFERENCES

Ada, N., Kazancoglu, Y., Sezer, M. D., Ede-Senturk, C., Ozer, I., & Ram, M. (2021). Analyzing barriers of circular food supply chains and proposing industry 4.0 solutions. *Sustainability*, *13*(12), 6812.

Aithal, P. S., & Sony, M. (2020). Design of Industry 4.0 readiness model for Indian Engineering Industry: Empirical Validation Using Grounded Theory Methodology. *International Journal of Applied Engineering and Management Letters (IJAEML)*, *4*(2), 124–137.

André, J. C. (2019). *Industry 4.0: Paradoxes and conflicts*. John Wiley & Sons.

Asghar, S., Rextina, G., Ahmed, T., & Tamimy, M. I. (2020). *The Fourth Industrial Revolution in the developing nations: Challenges and road map* (No. 102). Research Paper.

Bongomin, O., Nganyi, E. O., Abswaidi, M. R., Hitiyise, E., & Tumusiime, G. (2020). Sustainable and dynamic competitiveness towards technological leadership of industry 4.0: implications for East African community. *Journal of Engineering*, 2020, 1–22.

Chauhan, C., Singh, A., & Luthra, S. (2021). Barriers to industry 4.0 adoption and its performance implications: An empirical investigation of emerging economy. *Journal of Cleaner Production*, 285, 124809.

Chow, C. (2019). Karma Yoga: Application of Gita (2: 47) for Superior Business Performance During Industry 4.0. *Managing by the Bhagavad Gītā: Timeless Lessons for Today's Managers*, 103–135.

Çınar, Z. M., Zeeshan, Q., & Korhan, O. (2021). A framework for industry 4.0 readiness and maturity of smart manufacturing enterprises: a case study. *Sustainability*, 13(12), 6659.

Deloitte. (2018). Industry 4.0 and manufacturing ecosystems: Exploring the world of connected enterprises. Retrieved from: www2.deloitte.com/content/dam/Deloitte/in/Documents/manufacturing/in-manufacturing-industry-4-0-noexp.pdf

Dos Santos, L. M. A. L., da Costa, M. B., Kothe, J. V., Benitez, G. B., Schaefer, J. L., Baierle, I. C., & Nara, E. O. B. (2021). Industry 4.0 collaborative networks for industrial performance. *Journal of Manufacturing Technology Management*, 32(2), 245–265.

Dutta, G., Kumar, R., Sindhwani, R., & Singh, R. K. (2020). Digital transformation priorities of India's discrete manufacturing SMEs–a conceptual study in perspective of Industry 4.0. *Competitiveness Review: An International Business Journal*, 30(3), 289–314.

Fatorachian, H., & Kazemi, H. (2021). Impact of Industry 4.0 on supply chain performance. *Production Planning & Control*, 32(1), 63–81.

Goswami, M., & Daultani, Y. (2022). Make-in-India and industry 4.0: Technology readiness of select firms, barriers and socio-technical implications. *The TQM Journal*, 34(6), 1485–1505.

Government of India. (2019). Make in India. Retrieved from: www.makeinindia.com/home

Hajoary, P. K., & Akhilesh, K. B. (2021). Conceptual framework to assess the maturity and readiness towards Industry 4.0. In *Industry 4.0 and Advanced Manufacturing: Proceedings of I-4AM 2019* (pp. 13–23). Springer, Singapore.

Iyer, A. (2018). Moving from Industry 2.0 to Industry 4.0: A case study from India on leapfrogging in smart manufacturing. *Procedia Manufacturing*, 21, 663–670.

Jabbar, A., Abbasi, Q. H., Anjum, N., Kalsoom, T., Ramzan, N., Ahmed, S. & Ur Rehman, M. (2022). Millimeter-Wave Smart Antenna Solutions for URLLC in Industry 4.0 and Beyond. *Sensors*, 22(7), 2688.

Jahani, N., Sepehri, A., Vandchali, H. R., & Tirkolaee, E. B. (2021). Application of industry 4.0 in the procurement processes of supply chains: a systematic literature review. *Sustainability*, 13(14), 7520.

Jena, A., & Patel, S. K. (2022). Analysis and evaluation of Indian industrial system requirements and barriers affect during implementation of Industry 4.0 technologies. *The International Journal of Advanced Manufacturing Technology*, 120(3–4), 2109–2133.

Kadir, B. A., Broberg, O., & Souza da Conceição, C. (2018). Designing human-robot collaborations in industry 4.0: explorative case studies. In *DS 92: Proceedings of the DESIGN 2018 15th International Design Conference*, 601–610.

Kaliraj, P., & Devi, T. (Eds.). (2022). *Industry 4.0 Technologies for Education: Transformative Technologies and Applications*. CRC Press.

Kamble, S. S., Gunasekaran, A., & Sharma, R. (2018). Analysis of the driving and dependence power of barriers to adopt industry 4.0 in Indian manufacturing industry. *Computers in Industry*, 101, 107–119.

Kumar, P., & Tiwari, M. K. (2018). Lean six sigma 4.0: A conceptual framework for industry 4.0. *International Journal of Lean Six Sigma*, 9(2), 182–205.

Kumar, R., Singh, R. K., & Dwivedi, Y. K. (2020). Application of industry 4.0 technologies in SMEs for ethical and sustainable operations: Analysis of challenges. *Journal of cleaner production*, 275, 124063.

Kumar, S., Srivastava, M., & Prakash, V. (2023). Challenges and Opportunities for Mutual Fund Investment and the Role of Industry 4.0 to Recommend the Individual for Speculation. *New Horizons for Industry 4.0 in Modern Business*, 69–98.

Lambrechts, W., Sinha, S., & Marwala, T. (2021). BRICS and Industry 4.0. *The BRICS Order: Assertive or Complementing the West?*, 283–322.

Lödding, H., Riedel, R., Thoben, K. D., Von Cieminski, G., & Kiritsis, D. (Eds.). (2017). *Advances in Production Management Systems. The Path to Intelligent, Collaborative and Sustainable Manufacturing: IFIP WG 5.7 International Conference, APMS 2017, Hamburg, Germany, September 3–7, 2017, Proceedings, Part II*, 514. Springer.

Luthra, S., & Mangla, S. K. (2018). Evaluating challenges to Industry 4.0 initiatives for supply chain sustainability in emerging economies. *Process Safety and Environmental Protection*, 117, 168–179.

Mahindra & Mahindra. (2021). Industry 4.0 in Manufacturing. Retrieved from https://www.techmahindra.com/en-in/manufacturing/factory-of-future-auto/

Mavropoulos, A., & Nilsen, A. W. (2020). *Industry 4.0 and circular economy: Towards a wasteless future or a wasteful planet?*. John Wiley & Sons.

Muller, J. M., & Kazantsev, N. (Eds.). (2021). *Industry 4.0 in SMEs across the globe: Drivers, barriers, and opportunities*. CRC Press.

Nath, S. V., Dunkin, A., Chowdhary, M., & Patel, N. (2020). *Industrial Digital Transformation: Accelerate digital transformation with business optimization, AI, and Industry 4.0*. Packt Publishing.

Neville, J., & Doyle-Kent, M. (2022). The "Three I's" of Industry 4.0: A Framework for Irish Industry. *IFAC-PapersOnLine*, 55(39), 431–436.

Parhi, S., Joshi, K., Wuest, T., & Akarte, M. (2022). Factors affecting Industry 4.0 adoption–A hybrid SEM-ANN approach. *Computers & Industrial Engineering*, 168, 108062.

Primi, A., & Toselli, M. (2020). A global perspective on industry 4.0 and development: new gaps or opportunities to leapfrog?. *Journal of Economic Policy Reform*, 23(4), 371–389.

Rahimian, F. P., Goulding, J. S., Abrishami, S., Seyedzadeh, S., & Elghaish, F. (2021). *Industry 4.0 solutions for building design and construction: a paradigm of new opportunities*. Routledge.

Rao, P. V., & Roy, R. K. (2018). Industry 4.0 and Lean Six Sigma convergence: A case study of an automotive parts manufacturer. *International Journal of Productivity and Performance Management*, 67(9), 1595–1610.

Rockenschaub, T. (2016). *The Interrelationship of Industry 4.0 and Strategic Management-A Systematic Literature Review/Author Ing. Thomas Rockenschaub* (Doctoral dissertation, Universität Linz).

Safar, L., Sopko, J., Dancakova, D., & Woschank, M. (2020). Industry 4.0—Awareness in South India. *Sustainability*, 12(8), 3207.

Sarkar, B. D., Shankar, R., & Kar, A. K. (2022). Severity analysis and risk profiling of port logistics barriers in the Industry 4.0 era. *Benchmarking: An International Journal* (ahead-of-print).

Sarkar, B. D., Shankar, R., & Kar, A. K. (2023). Port logistic issues and challenges in the Industry 4.0 era for emerging economies: an India perspective. *Benchmarking: An International Journal*, 30(1), 50–74.

Saxena, A., Singh, R., Gehlot, A., Akram, S. V., Twala, B., Singh, A., & Priyadarshi, N. (2022). Technologies Empowered Environmental, Social, and Governance (ESG): An Industry 4.0 Landscape. *Sustainability*, 15(1), 309.

Seth, D., Gupta, M., & Singh, B. J. (2022a). A Study to Analyse the Impact of Using the Metaverse in the Banking Industry to Augment Performance in a Competitive Environment. In *Applying Metalytics to Measure Customer Experience in the Metaverse*, 9–16. IGI Global.

Seth, D., Gupta, M., & Singh, B. J. (2022b). Study in the BFSI Sector on the Role of AI in Human Resource Management. In *Applying Metalytics to Measure Customer Experience in the Metaverse*, 173–181. IGI Global.

Shanker, K., Shankar, R., & Sindhwani, R. (2019). Advances in industrial and production engineering. *Select proceedings of FLAME 2018 book series*.

Sharma, A., & Singh, B. J. (2022). Understanding LSS 4.0 through golden circle model and reviewing its scope in Indian textile industry. *International Journal of Six Sigma and Competitive Advantage*, *14*(1), 120–137.

Singh, M., & Rathi, R. (2019). A structured review of Lean Six Sigma in various industrial sectors. *International Journal of Lean Six Sigma*, *10*(2), 622–664.

Singh, M., Rathi, R., Khanduja, D., Phull, G. S., & Kaswan, M. S. (2020). Six Sigma methodology and implementation in Indian context: a review-based study. *Advances in Intelligent Manufacturing: Select Proceedings of ICFMMP 2019*, 1–16.

Siqueira, F., & Davis, J. G. (2021). Service computing for industry 4.0: State of the art, challenges, and research opportunities. *ACM Computing Surveys (CSUR)*, *54*(9), 1–38.

Sodhi, H. S. (2021). An evaluation of sustainability of Industry 4.0 aspects. *International Journal of Productivity and Quality Management*, *34*(3), 399–414.

Sony, M., & Aithal, P. S. (2020). Developing an industry 4.0 readiness model for Indian engineering industries. *International Journal of Management, Technology, and Social Sciences (IJMTS)*, *5*(2), 141–153.

Sun, X., Yu, H., & Solvang, W. D. (2022). Towards the smart and sustainable transformation of Reverse Logistics 4.0: a conceptualization and research agenda. *Environmental Science and Pollution Research*, *29*(46), 69275–69293.

Suresh, N., Hemamala, K., & Ashok, N. (2018). Challenges in implementing industry revolution 4.0 in INDIAN manufacturing SMES: insights from five case studies. *International Journal of Engineering & Technology*, *7*(2.4), 136–139.

Tripathi, S., & Gupta, M. (2021). Indian supply chain ecosystem readiness assessment for Industry 4.0. *International Journal of Emerging Markets* (ahead-of-print).

Ustundag, A., & Cevikcan, E. (2018). *Industry 4.0: managing the digital transformation*. Springer Nature.

Yoon, K. P., Lee, Y. H., & Kim, H. J. (2019). Lean Six Sigma 4.0: An advanced model of quality improvement in the steelmaking process. *Quality Innovation Prosperity*, *23*(2), 1–18.

10 Role of Lean Six Sigma 4.0 and Digital Forensic Tools in Cyber Investigation

Tejasvi Pandey, Varun Vevek, and Astha Dhall

10.1 INTRODUCTION

With the evolution of digital media, there has been an exponential rise in its users, accompanied by their data. As a result, many applications and pieces of software have emerged which enhance our day-to-day lifestyle and have become part of our modern-day life [1]. These are the platforms where businesses, commerce, and communication technology exist. This synergy of the physical world with the Internet of Things (IoT) and Information Technology (IT) has led to numerous benefits and become the backbone of the 21st century. The scenario pertaining since COVID in 2020 has seen a decreased number of physical crimes and has caused an alarming increase in cybercrimes [2]. At the same time, it is accompanied by a new arena of cybercrimes, ranging from hacking bank servers and data leakage to identity theft and cyberbullying, all of which can be commissioned from a remote location. These crimes have a significant socio-economic impact on the global economy and individual lives. Therefore, it is imperative to identify, apprehend, and convict the perpetrators [3].

Cyber Forensics deals with digital or cyber-based crimes in which victims have fallen victim in either an app or software. Alternatively, others are abused or harassed online, including by sharing data. By using cyber forensic tools, it is possible to probe the evidence. Unfortunately, there are certain limitations to investigating cybercrimes [4]. The crimes are often committed from different territories, requiring investigation across international borders, and are subject to different laws and legal systems. Furthermore, they are accompanied by a large and complex structure of software frameworks and vast volumes of data [5]. Therefore, it becomes essential to incorporate an advanced continuous improvement approach like Lean Six Sigma (LSS) and the LSS 4.0 approach, with their tools and techniques, into digital forensics [6].

The integration of LSS and Internet of Things-based tools provides a digital approach when we talk about LSS 4.0. The goal of LSS 4.0 is to cut down on waste and defects (variation) while simultaneously improving efficiency and output with

 DOI: 10.1201/9781003381600-10

IoT [7]. Simple LSS approaches have become more technologically capable as manufacturing transforms into a more digitized environment with the development of Industry 4.0 (I4.0) technologies [8]. Consequently, the development of new I4.0 technologies like the Internet of Things (IoT), big data and data analytics, as well as augmented and virtual reality, may lead to the emergence of the fourth revolution in LSS, or LSS 4.0 [9]. When used together, I4.0 and LSS 4.0 are much more effective than when one procedure is carried out autonomously of the other. In addition, a thorough understanding of the functions, duties, and frameworks of the employee, as well as the prerequisites for their training, are essential to the achievement of an effective LSS 4.0 adoption in forensics.

This review discusses the advances in digital forensics over time, its relevance within LSS 4.0 strategy in the current context, the possible development areas, how digital forensics could grow quickly, and its application.

10.2 DIGITAL FORENSICS

Digital forensics is a branch of forensic science focusing on recovering and analyzing digital evidence. First, materials found inside systems are further analyzed to find the perpetrator. Then, evidence is identified, preserved, and analyzed according to the laws and regulations of the territory. Digital forensics is further categorized depending upon specializations and functionalities that are shown in Figure 10.1 and described in detail below.

> *Database Forensics*: Database forensic study of databases involves analyzing all the files and their primary and sensitive contents (metadata) stored in RAMs, files, hard drives, etc. This aims to monitor unapproved admittance to control data and likewise notice strange conduct around the information [10].
> *Mobile Device Forensics*: This focuses on the discovery, recovery, and analysis of evidence from mobile devices. It can entail collecting call records and text messages from all such applications present in the device, emails, internet history, social media applications, etc, [11].
> *Computer Forensics*: This deals with finding, recovering, and analyzing evidence from computer devices suspected of being used in the commission of, or related to, the crime. It contains forensic tools associated with data analysis, MAC-OS analysis, and mobile device tools [12]. In addition, it examines system file logs,

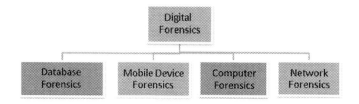

FIGURE 10.1 Classification of digital forensics.

records deleted files, discovers hidden files, and analyzes web history, emails, documents, etc.

Network Forensics: Network Forensics analyzes incoming and outgoing data packets between networks used for communication. It is used for authentication and intrusion detection by analyzing the source and destination data [13].

10.3   LEAN SIX SIGMA 4.0

LSS as a continuous improvement approach has been implemented in numerous manufacturing and service organizations over the globe [14]. This approach is able to reduce the non-value-added activities and variability in the production line to obtain superior quality at optimum cost. The LSS approach is primarily executed through the DMAIC (Define-Measure-Analyze-Improve-Control) methodology in most projects. The literature also reveals that the DMAIC approach was implemented for continuous improvement in process industries [15], manufacturing industry [16], printing industry [17], healthcare industries [18], etc. In recent years, practitioners have increasingly focused on integrating advanced technologies like Industry 4.0 with continuous improvement methods and redesigning their processes to improve operational performance [19]. LSS 2.0 and 3.0 revealed that an improvement system is one that "can produce, sustain, and successfully integrate improvements in any environment, culture, and business" [20]. Consequently, the development of new I4.0 technologies like the internet of things (IoT), big data and data analytics, as well as augmented and virtual reality, may lead to the emergence of the fourth revolution in LSS, known as LSS 4.0 [21].

10.4   PROCEDURE TO TRANSFORM CONVENTIONAL FORENSICS INTO A DIGITAL SCENARIO

A. The data either needs to be retrieved or extracted from the device.

B. The software being used in the device needs to be identified.

C. In the case of desktop/monitor, the installed windows operating system needs to be identified.

D. In mobile phones, determine whether Android, iOS, or windows is being used.

E. Decide whether the data needs to be recovered from the phone's hardware, i.e. SD card, or the software, or over the network address of the respective person.

F. In the case of a desktop, the data needs to be recovered over a network or retrieved from the CPU.

G. If a cyber-attack is over the Internet, it needs to be determined whether it was an email attack, link generated, or phishing over an I-frame, etc.

H. The cyber counterattack needs to be performed with the help of specific tools.

I. The tool to be selected must be compatible with the operating system of the electronic evidence to be examined.

J. The tool must always be used on the copy of the data if already retrieved.

K. In the case of retrieving or analyzing the data online, each tool applied and the steps taken must be noted for the chain of custody integrity.

L. Because digital evidence may be readily affected by powerful magnetic fields which can influence data integrity in particular storage mediums, the integrity of the targeted evidence must be considered.

M. The validation testing of the tools must be done before applying them to the scan.

10.4.1 STANDARD OPERATING PROTOCOL (SOP) OF INVESTIGATION

The Standard Operating Protocol (SOP) used to conduct the analysis must maintain the credibility and integrity of the evidence [22]. The SOP is designed to:

*   *Protect*: The computer system must be protected from manipulation, damage, data corruption, and virus infection in general.
*   *Discover*: All files in the subject system, including existing regular files, deleted files that persist, hidden files, password-protected files, and encrypted data, must be discovered.
*   *Recover*: To the extent feasible, all files detected must be recovered..
*   *Reveal*: To the greatest extent practicable, expose all hidden and temporary files utilized by the operating system and application programs.
*   *Access*: If possible and legal, try to access all of the content of protected or hidden files.
*   *Analyze*: Any essential information discovered in specific sections of the disk.
*   *Layout*: An overall analysis of the concerned computer system or device is conducted. This analysis documents all the files and data available and discovered in the system. This layout also includes a system overview, file information, etc. Any attempts to tamper with the files, for example, add, delete, modify and encrypt, will also be revealed in the report.
*   *Provide Expert Opinion*: A forensic expert who has performed an analysis must testify and give opinions regarding the analysis.

10.5 ROLE OF LSS 4.0 IN FORENSIC DNA

The efficiency of casework can be increased by gathering baseline data, by having an understanding of the current situation and processes, and by providing data on process mapping. Improving administrative procedures, personnel management, hardware, software, and resource usage also provide effective management of case files and performance monitoring of all phases of analysis [23].

It is vital to deploy distributed digital forensic (DDF) techniques and demonstrate their success in real world scenarios where the limitations of conventional investigation have been reached. It is also important to establish a set of system specifications for DDF software and suggest a compact architecture to meet those requirements. A digital forensics toolkit (DDF) is not just a generic framework, but rather a targeted solution that can be optimized for some specific purpose. It should be able to utilize anywhere between tens and hundreds of machines across a fast LAN and should speed up execution time linearly. A distributed forensic investigation toolkit (DDF) should be simple to install and use by an end user and should make few assumptions about the underlying infrastructure. It should also make sure that developing new processing functions for the DDF toolkit requires the same amount of time and expertise as developing them for the sequential case. [24].

Digital forensic labs that are owned or operated by states are forced to cut back on the quantity of data they forensically examine or else risk growing workload backlogs. Managers and employees of digital forensic laboratories now have more tactical options due to the introduction of enhanced previewing. Algorithms can be written that will result in a disk being imaged automatically to a server without requiring human input. Digital forensic labs are also able to create their own enhanced previewing tools using open-source technologies which can be deployed easily at very low cost [26].

10.5.1 DMAIC Process in Digital Forensics

Lean Six Sigma uses the DMAIC process, which stands for Define, Measure, Analyze, Improve, and Control. The DMAIC process is a five-step approach used to identify and solve problems within a process. In the Define step, the problem or opportunity is defined, and the goals and objectives of the project are established. This step involves identifying the specific problem or opportunity that needs to be addressed, defining the project's scope and the stakeholders involved. In digital forensics, this phase could involve identifying a problem or opportunity related to digital evidence collection, analysis, or preservation and establishing a project team to address it. The next step is Measure. Here, the data is collected and analyzed to measure the current performance of the process [27–29]. This step involves identifying the key performance indicators (KPIs) used to measure the process and collecting data to establish a performance baseline. In digital forensics, this phase could involve collecting data on the current time and costs associated with collecting and analyzing digital evidence, as well as the accuracy and completeness of the evidence [30]. In the Analyze step, the data collected in the Measure step is analyzed to identify the root cause of the problem or opportunity. This step involves using statistical techniques and process mapping to identify areas for improvement. In digital forensics, this phase could involve analyzing the data to identify bottlenecks in the digital evidence collection and analysis process, and to identify sources of variability in the accuracy and completeness of the evidence [31]. The Improve step is used to develop and implement solutions to improve the digital forensic process. This includes using advanced technologies such as IoT, AI, and big data analytics to automate repetitive tasks, eliminate bottlenecks, and improve the accuracy and speed of the process [32]. This step also includes developing a detailed implementation plan, which outlines how the solutions will be implemented and who will be responsible for implementing them. In digital forensics, this phase could involve implementing new technologies or procedures to automate the collection and analysis of digital evidence, or implementing a new system for storing and preserving digital evidence [33]. Finally, in the Control step, the digital forensic process is monitored and controlled to ensure that the improvements made in the

Improve phase are sustained. This includes implementing standardized procedures, using control charts to monitor the process, and conducting regular audits to ensure compliance with the new process. This phase also includes developing a control plan, which outlines how the process will be monitored and controlled, who will be responsible for monitoring and controlling the process, and how the process will be audited. In digital forensics, this phase could involve implementing a new system for monitoring the process, such as a dashboard, and conducting regular audits to ensure compliance with the new process and identify any areas that need further improvement.

10.6   RECOMMENDATIONS FOR FUTURE DIRECTIONS

1.   Databases should be created for every division in a forensic science laboratory to reduce costs, maximize the process of evidence gathering and to speed up the disposal of cases.
2.   Investigation officers and police personnel should be trained by professionals for collecting evidence at the crime scene which would lead to least contamination of the evidence and would yield better results upon examination.
3.   A Lean Six Sigma, DMAIC roadmap should be applied in every step to develop a process with greater efficiency. It should also be applied in Forensic science laboratories to:
     a.   Reduce over-production of chemicals.
     b.   Use proper instruction which will minimize contamination and error rates.
     c.   Minimize human and machine error rates.
     d.   Prepare a base line for comparison purposes.
     e.   Keep a track of tools and techniques which give efficient results within a limited time period and reduce non-value-added steps in between.

10.7   CONCLUSION

As our current world is transforming into a cyber world, it has provided great improvements in conducting day-to-day activities. This opened many opportunities for innovation and brought a surge of cybercrimes, ranging from data theft to financial fraud and malware attacks. This surge demands further advances in digital and cyber forensics investigation: faster and more effective incident response, forensic tools with heterogeneous applications, standardized regulations for investigations, etc, through a LSS 4.0 approach. We have discussed significant challenges that the digital forensics community faces and some measures to resolve them. There has been an increase in measures and software which provide better security. Practical tools should be made either from pre-existing tools discussed in the above sections or new tools that counter these anti-forensic methodologies while invading privacy as little as possible. People in today's modern world should have enough knowledge to protect themselves from cyberattacks.

REFERENCES

[1] Marshall. (2021). Digital forensic tool verification: An evaluation of options for establishing trustworthiness.– ScienceDirect. Retrieved January 10, 2023, from www.sciencedirect.com/science/article/abs/pii/S2666281721000895

[2] Luciano, Baggili, Topor, Casey, & Breitinger. (2018). *Digital Forensics in the Next Five Years*, ACM Digital Library. Retrieved January 10, 2023, from https://dl.acm.org/doi/10.1145/3230833.3232813

[3] Cameron. (n.d.). Digital Forensics: 6 Security Challenges | IEEE Computer Society. Retrieved January 10, 2023, from www.computer.org/publications/tech-news/research/digital-forensics-security-challenges- cybercrime

[4] Caviglione, Wendzel, & Mazurczyk. (2017). *The Future of Digital Forensics: Challenges and the Road Ahead CSDL | IEEE Computer Society*. Retrieved January 10, 2023, from www.computer.org/csdl/magazine/sp/2017/06/msp2017060012/13rRUwghd7E

[5] Arthur, & Venter. (2004). *An Investigation Into Computer Forensic Tools*. In *Proceedings of the {ISSA} 2004 Enabling Tomorrow Conference, 30 June- 1 July 2004, Gallagher Estate, Midrand, South Africa* (pp. 1–11). ISSA. http://icsa.cs.up.ac.za/issa/2004/Proceedings/Full/060.pdf

[6] Devendran, Shahriar, & Clincy. (2015). *A Comparative Study of Email Forensic Tools. Journal of Information Security*, 6(2) pp. 111–117. https://doi.org/10.4236/jis.2015.62012

[7] Santhi, Kanakam, & Hussain. (2017). *Cyber Forensic Science to Diagnose Digital Crimes- A study. International Journal of Computer Trends and Technology*, 50(2), 107–113. https://doi.org/10.14445/22312803/ijctt-v50p119

[8] Gonzales, Schofield, & Hagy. (2012). *Investigations Involving the Internet and Computer Networks*. In *National Institute of Justice*. U.S. Department of Justice. www.ojp.gov/pdffiles1/nij/210798.pdf

[9] Ashcroft, Daniels, & Hart. (2014). *Forensic Examination of Digital Evidence: a Guide for Law Enforcement*. In *National Institute of Justice*. U.S. Department of Justice. www.ojp.gov/pdffiles1/nij/199408.pdf

[10] Ogden. (2017). *Mobile Device Forensics: Beyond Call Logs and Text Messages*. Crime Scene Investigator Network. Retrieved January 10, 2023, from www.crime-scene-investigator.net/mobile-device-forensics-beyond-call-logs-and-text- messages.html

[11] Gonzales, Schofield, & Hagy. (2007). *Digital Evidence in the Courtroom: A Guide For Law Enforcement and Prosecutors | Office of Justice Programs*. US Department of Justice. Retrieved January 10, 2023, from www.ojp.gov/ncjrs/virtual-library/abstracts/digital-evidence- courtroom-guide-law-enforcement-and-prosecutors

[12] Holder Jr., Robinson, & Rose. (2009). *Electronic Crime Scene Investigation: An On-the-Scene Reference for First Responders*. US Department of Justice. Retrieved January 10, 2023, from https://nij.ojp.gov/library/publications/electronic-crime-scene-investigation-scene- reference-first-responders

[13] Carroll, Brannon, & Song. (2017). Computer Forensics: Digital Forensic Analysis Methodology. Crime Scene Investigator Network. Retrieved January 10, 2023, from www.crime-scene-investigator.net/computer-forensics-digital-forensic-analysis-methodology.html

[14] Carroll. (2017). *Challenges in Modern Digital Investigative Analysis*. Crime Scene Investigator Network. Retrieved 26 Apr. 2019, www.crime-scene-investigator.net/challenges-in-modern-digital-investigative-analysis.html.

[15] Lillis, Becker, O'Sullivan, & Scanlon. (2016). *Current Challenges and Future Research Areas for Digital Forensic Investigation*. arXiv.org, Cornell University. Retrieved January 10, 2023, from https://arxiv.org/abs/1604.03850v1

[16] Kapoor, Taneja, & Kumar. (2019). *Digital Forenisc Tools*. International Journal of Engineering and Advanced Technology (IJEAT), 9(25). https://doi.org/10.35940/ijeat. B3980.129219

[17] Dezfouli, & Dehghantanha. (2014). *Digital forensics trends and future*. International Journal of Cyber-Security and Digital Forensics (IJCSDF), 3(4). http://usir.salford. ac.uk/id/eprint/34014/

[18] Yadav, Ahmad, Shekhar. (2011). *Analysis of Digital Forensic Tools and Investigation Process*. In: Mantri, A., Nandi, S., Kumar, G., Kumar, S. (eds) High Performance Architecture and Grid Computing. HPAGC 2011. *Communications in Computer and Information Science, 169*. Springer, Berlin, Heidelberg. https://doi.org/10.1007/978-3-642-22577-2_59

[19] Garfinkel. (2010) *Digital forensics research: The next 10 years. Digital Investigation, 7*, Supplement. ISSN 1742-2876. https://doi.org/10.1016/j.diin.2010.05.009. www. sciencedirect.com/science/article/pii/S1742287610000368

[20] Ghazinour, Vakharia, Kannaji & Satyakumar. (2018). *A study on digital forensic tools. 2017 IEEE International Conference on Power, Control, Signals and Instrumentation Engineering (ICPCSI)*. https://doi.org/10.1109/ICPCSI.2017.8392304.

[21] Guo, Slay, Beckett. (2009). *Validation and verification of computer forensic software tools—Searching Function. Digital Investigation, 6* Supplement. ISSN 1742-2876. https://doi.org/10.1016/j.diin.2009.06.015.

[22] Arthur, & Venter. (2004). *An Investigation Into Computer Forensic Tools*. In Issa, 1–11.

[23] (2019). List of 15 Most Powerful Forensic Tools. *IFF Lab*. www.ifflab.org/list-of-15-most- powerful-forensic-tools/.

[24] Tiwari. (2021). *Twitter Loses Its Intermediary Status in India, Here Is What It Means*. India Today. www.indiatoday.in/technology/features/story/twitter-loses-its-intermedi ary-status-in- india-here-is-what-it-means-1815491-2021-06-16.

[25] Agnihotri. (2015). *Forensic Data Extraction- Bulk Extractor. Forensic Data Extraction- Bulk Extractor*. www.linkedin.com/pulse/forensic-data-extraction-bulk-extractor-mayur- agnihotri.

[26] Hakmeh, Taylor, Peters & Ignatidou. (2021) *The COVID-19 Pandemic and Trends in Technology – Transformations in Governance and Society*. Chatham House. www.chathamhouse.org/2021/02/covid-   19-pandemic-and-trends-technology/ 04-infodemic-and-covid-19-disinformation.

[27] Ani Petrosyan (2023). Most reported types of cyber-crime worldwide 2022. Statista. www.statista.com/statistics/184083/commonly-reported-types-of- cyber-crime.

[28] Alfaro, Madrigal & Hernandez. (2020). *Improving Forensic Processes Performance: A Lean Six Sigma Approach. Forensic Science International: Synergy, 2*. https://doi.org/ 10.1016/j.fsisyn.2020.02.001.

[29] Smętkowska & Mrugalska. (2018) *Using Six Sigma DMAIC to Improve the Quality of the Production Process: A Case Study. Procedia – Social and Behavioral Sciences, 238*. https://doi.org/10.1016/j.sbspro.2018.04.039.

[30] Hill, Thomas, Mason-Jones & El-Kateb. (2017). *The Implementation of a Lean Six Sigma Framework to Enhance Operational Performance in an MRO Facility. Production & Manufacturing Research, 6* (1). https://doi.org/10.1080/21693 277.2017.1417179.

[31] Richard, & Kupferschmid. (2011). *Increasing Efficiency of Forensic DNA Casework Using Lean Six Sigma Tools*. Office of Justice Programs www.ojp.gov/pdffiles1/nij/gra nts/235190.pdf.

[32] Chiarini, & Kumar. (2021). *Lean Six Sigma and Industry 4.0 Integration for Operational Excellence: Evidence From Italian Manufacturing Companies. Production Planning & Control, 32*(13). https://doi.org/10.1080/09537287.2020.1784485.

[33] Lameijer, Pereira, & Antony. (2021). *The Implementation of Lean Six Sigma for Operational Excellence in Digital Emerging Technology Companies. Journal of Manufacturing Technology Management, 32*(9). https://doi.org/10.1108/ jmtm-09-2020-0373.

Index

Note: Page numbers referring to figures are in **bold** and those referring to tables are in *italic*.

Printed in the United States
by Baker & Taylor Publisher Services